W9-BVB-783

You Had Me at Pét-Nat

You Had Me at Pét-Nat

A Natural Wine-Soaked Memoir

Rachel Signer

New York

This is a work of memoir. Events and dialogue have been reconstructed to the best of the author's ability, through memory and notes, as well as interviews and research.

Cover design and illustration by Kimberly Glyder

Hachette Books
Hachette Book Group
1290 Avenue of the Americas
New York, NY 10104
HachetteBooks.com
Twitter.com/HachetteBooks
Instagram.com/HachetteBooks

First Edition: October 2021

Published by Hachette Books, an imprint of Perseus Books, LLC, a subsidiary of Hachette Book Group, Inc. The Hachette Books name and logo is a trademark of the Hachette Book Group.

The Hachette Speakers Bureau provides a wide range of authors for speaking events.

To find out more, go to www.hachettespeakersbureau.com or call (866) 376-6591.

The publisher is not responsible for websites (or their content) that are not owned by the publisher.

Print book interior design by Amy Quinn.

Library of Congress Cataloging-in-Publication Data

Names: Signer, Rachel, author.
Title: You had me at pet-nat : a natural wine-soaked memoir / Rachel Signer.
Other titles: You had me at pétillant-naturel
Description: First edition. | New York : Hachette Books, 2021.
Identifiers: LCCN 2021007187 | ISBN 9780306924743 (hardcover) | ISBN 9780306924750 (ebook)
Subjects: LCSH: Signer, Rachel. | Vintners--Australia--Biography.
Classification: LCC TP547.S44 A3 2021 | DDC 663/.20092 [B]--dc23
LC record available at https://lccn.loc.gov/2021007187ISBNs: 978-0-306-92474-3 (hardcover), 978-0-306-92475-0 (ebook)

Printed in the United States of America

LSC-C

Printing 1, 2021

For Simone

In Europe then we thought of wine as something as healthy and normal as food and also as a great giver of happiness and well-being and delight. Drinking wine was not a snobbism nor a sign of sophistication nor a cult; it was as natural as eating and to me as necessary, and I would not have thought of eating a meal without drinking either wine or cider or beer.

—Ernest Hemingway, *A Moveable Feast*

Things had reduced themselves to a tragicomic scenario; on the one hand, the man identifying the woman as an angel, on the other, the angel identifying love as something only a little short of pathology.

—Alain de Botton, *Essays in Love*

This book was largely inspired by and written on Peramangk Country. I pay respect to the Traditional Owners and ongoing custodians of this land, and to their Elders, past and present.

Contents

Preface

2018

IN WINEMAKING, A SUPPOSEDLY SIMPLE TASK IS GENERALLY AN EXCELlent opportunity to make a fool of oneself.

A soft wind rippled through the warm March afternoon. Overhead, a pair of black cockatoos squawked loudly, careening over the valley behind the winery. Shuffling around in my newly purchased waterproof Rossi boots, made in South Australia, I rearranged my stance as I prepared to transfer the juice from a barrel to a vat full of spent grape skins. We were making *piquette*, a low-alcohol wine, traditionally served to vineyard workers in France, derived from already pressed grapes topped up with fresh juice and water.

Maybe like this? I held the hose in one hand, my legs wide apart. Wildman always did it so quickly, I hardly had a chance to observe him. And I hadn't witnessed this technique in France, when I'd had my stint picking grapes at Domaine Mosse last year. Most wineries would simply use a mechanical pump to transfer fresh juice or finished wine from one vessel to another. Nearly every winery in the world would do it that way. But here at Lucy Margaux, we did things differently. Pumps, for whatever reason, were against Wildman's ideology. Too aggressive on the wine, he said, succinct as always in his explanations.

Leaning down, I placed my lips over the hose opening and inhaled; my chest tightened. Wasn't siphoning a basic survival skill? Shouldn't I have learned it in Girl Scouts, back in Virginia, instead of making s'mores all those years? Squatting to lower the hose to allow the juice to flow into the bucket, I glimpsed the pale liquid filling the part around the barrel opening. It was coming! I lowered and sucked, lowered and sucked, trying to copy the maneuver I'd seen Wildman do with complete fluency.

"Heeeeeuuuuuuuuuh!" I inhaled with all the lung capacity I possessed, trying to work with gravity. Gravity, what did I know about gravity, anyway? I, who only took physics in university because it was a required course. Who wrote a final paper about snowboarding because I knew it would just pass me. But surely I could do this task—couldn't I?

As a freelance journalist who had spent years writing about wine, I thought I had an idea of how it was made. Countless press trips and tasting seminars had given me a strong sense of the basics: pick grapes, destem or not, jump on them or ferment "carbonically," macerate grapes or press directly, do punchdowns, transfer free-run juice into barrels or tanks, press the grapes again and again and again. But my two-week experience in France the previous year had shown me the reality of this work—I saw how much logistical coordination, technical prowess, and intuitive knowledge of fermentation went into making a stable, drinkable wine. Especially given that these were natural wines—free of any corrective additives and free of or very low in preservatives—there was no room for error.

Now, I had somehow waded back into the natural winemaking trenches, and the stakes were different. That was partly because of my intimacy with the winemaker, and partly because I was making five barrels of my own wine, plus a small batch of *pétillant-naturel*, a rustic sparkling wine made without any added yeasts or sugar (a.k.a. "pét-nat"). This was not an internship. This was, quite suddenly and surprisingly, my life. Nothing I'd experienced in my eight years of living in New York as an academic, writer, or wine salesperson could have prepared me for making natural wine in Australia.

I threw down the hose and marched into the winery with an air of exasperation. Wildman was directing the interns in sniffing the ferments, looking for potential problems. "Look for Samboy chips," said Wildman. "That's volatile acidity."

"Uh, what are those—Samboy chips?" Sev, who was originally from France and managed one of my favorite Manhattan wine bars, the Ten Bells, asked Wildman what we were all wondering.

"Vinegar chips. I'll buy you a pack so you know the smell," Wildman replied, standing in the center of the massive shed, while the others hovered over red plastic tubs of fermenting grapes, which we had picked that week. "And look for banana and nail polish, that's ethyl acetate. And reduction—match sticks."

Wildman walked purposefully toward one vat, where Raphael stood wearing a concerned expression. Raph, who was staying with us for a month from London, where he'd established himself as one of the city's best sommeliers, said he wasn't sure if the grapes smelled all right. Never mind my siphoning task, Wildman needed to be absolutely sure that every batch of fermenting grapes smelled good. Without the option of correcting the wines using something purchased from an oenological warehouse, nothing was more important than ensuring healthy ferments.

Since the Lucy Margaux winery was a sulfur-free zone—the extreme form of natural winemaking—we had to take utmost care to protect each vessel of fermenting wine, or else it could be ruined overnight. This would mean thousands of dollars lost and countless hours of work wasted—not just the picking but the careful pruning and cultivating that the growers, from whom Wildman purchased the grapes, had done throughout the year. Sulfur additions are the most controversial topic in the realm of natural winemaking; the mere mention of sulfur can provoke eye-rolling, sneers, and snide remarks or even insults from both sides of the argument—some people believe sulfur is necessary and prefer it be added, while others completely vilify the preservative.

But making wine without adding sulfur can be very tricky. Through our sniffing strategy, we were desperately trying to avoid potential faults such as "volatile acidity" and "mouse," both of which could render the wine faulty and essentially undrinkable.

As of five years earlier, I had never heard of natural wine, this strange category of alcoholic beverage, when I took a job that changed my life. I was in my late twenties, recovering from an intense period as an academic, looking for love despite a persistent incapability to value myself, and generally lost. All I knew was that I wanted to be a writer. And that I could continue surviving in New York by waiting tables. Then a particular rosé pét-nat from France set off a wild woman inside me who, though I didn't know it at the time, was dying for a new existence. And I got one, but not at all in the way I expected.

— One —

Shift Drinks

2013–2014

"Excuse me?" The tone suggested an apartment in Chelsea, career in advertising, perfectly touched-up highlights, maybe three-inch Louboutins if I looked under the table. I stopped in my tracks, instinctively adjusting my apron, and turned my head of unwashed hair toward the woman.

"Could I have some ketchup with my burger, please? Thanks so much." She smiled calmly, and I managed to mask my grimace. This was the fifteenth time on my brunch shift that I'd been sent to traverse the busy front dining room, then walk past the second dining area and into the kitchen, all to pick up a single ramekin of ketchup and deliver it to a table.

It was to be expected that brunch diners, especially those having a burger, would request ketchup. Our burger, however, came with a side of aioli, and "we" (the kitchen) strongly preferred that guests enjoy it with that mayonnaise-based topping—the European way. The restaurant stood proudly in a former textile factory converted to the very first boutique hotel in Brooklyn. To the remarks whispered in the grungy streets of Williamsburg, the surrounding neighborhood, that we were too snobby for our own good, we nodded in silent agreement. Indeed, most of our guests were posh Manhattanites. During my training week, I had somehow been allowed to wait on Diane Keaton herself, who had batted her eyes at me, crossed her pantsuited legs, and asked for "Cabernet on ice." I nearly tripped over my own feet while showing her a bottle from Southern France, a blend that contained Cabernet Sauvignon, which she refused to taste, waving a hand and insisting, "Oh, I'm sure it's fine." A Reynard waiter would have normally sneered—"Cabernet on ice" was not how we did wine. But Diane could have ordered a Long Island iced tea, and we would have obliged.

Reynard was an anomaly in that there was no Prosecco served at brunch—mimosas were prepared with organic French Crémant, made in the elaborate double-fermentation style of Champagne. This was considered more natural, and therefore superior to the Charmat method that most Proseccos endure, where sugar, yeast, and mostly dry wine are mixed together in a tank to produce bubbles. The wine list was strictly French *vin nature*, sourced via four or five small import companies who visited their producers regularly and therefore could vouch for their farming and vinification approaches. Our by-the-glass selection never included Merlot and only occasionally Chardonnay, but it often featured Gamay, Cabernet Franc, and the oddity known as "Muscadet"—a mineral-tasting white wine made of the Melon de Bourgogne grape, from vineyards along France's Atlantic coast. We abstained from oaky, rich wine, such as one would find in certain Bordeaux châteaux or around Napa Valley. It was no accident that the varieties and producers of the Loire Valley were prominent at Reynard—the wine director, Lee Campbell, had previously been a sales rep for Louis/Dressner Selections, one of the original New York natural wine importers with a portfolio steeped in Loire Valley wines.

As I grabbed the ketchup, I cast a quick glance toward the line cooks. Reynard had installed an "open kitchen," a novelty when the restaurant opened in 2013, which meant that back- and front-of-house could easily interact and cooks didn't feel like they were held in cages, away from guests. It was meant to be more civilized and democratic—with the bonus that it promoted a fantastic level of flirtation between line cooks and servers, me included.

I caught Chet's eye—he was the one cook who set my heart aflutter—and he winked. In response, my cheeks reddened almost as much as his, which were warm from the grill he'd been standing over, flipping and poaching eggs all day. This little ritual had been going on for several weeks, and we often got "in trouble" with the managers for chatting and giggling when I stopped in the kitchen to pick up dishes. Our adolescent interactions got me through the long shifts. I sometimes began work as early as 6 a.m. if my shift included breakfast as well as lunch or brunch. That morning, before leaving my apartment in Bed-Stuy, I'd swiped on some red lipstick to go with my tousled bed head, frumpy sweater, and leggings. Now I felt that this small burst of color had done the trick.

I delivered the ketchup to the woman eating the burger, cleared some plates, and went over to the bar to run a credit card. There, the red-headed bartender, Jarrett, who when he wasn't slinging mimosas and glasses of Cabernet Franc worked as a comedian, sidled over and nudged my elbow. I looked down: he was holding a shot of something brown. I inhaled the strong, medicinal fragrance of Fernet-Branca. He had one for himself in the other hand, and we quickly clinked our glasses and threw them back before anyone nearby could notice. It was a normal part of weekend brunch. Without those shots, we floor staff might have finally lost our shit at some table that made too many annoying requests.

Finally, it was time for me to toss my apron into the bin and have a seat at the bar for my shift drinks—plural. Thanks to the generosity of the owner, pioneering farm-to-table restaurateur Andrew Tarlow, we were permitted two on-the-house drinks after work at Reynard.

"What'll it be?" Jarrett smiled at me. He was in a much better mood than I was—our shifts were the same length, but he earned probably $200 more than me.

"Obviously, I'll have the 'Pièges à Filles,'" I told him brightly. Jarrett began pouring the bubbly, pink liquid—whose name translated, literally, as "girl trap"—into a wine glass while I watched eagerly. I had recently discovered this concoction. A bit of internet research had told me it was made in the Loire Valley from the red grape Gamay, by a pair of *négociant* winemakers—meaning they purchased grapes rather than growing them. Oddly, this duo made exclusively *pétillant-naturel*, or pét-nat for short.

I was familiar with Gamay from another wine we served by the glass, made by Olivier Cousin. According to a fellow server named Trevor, who was quite an authority on French wine, this Olivier Cousin had been heavily fined by the French authorities for the seemingly mundane act of writing the grape variety on his bottle labels. Clearly, these so-called *natural* winemakers were subversive. That was interesting to me, but more exciting was the way the wines tasted. Olivier Cousin's Gamay met my palate like salty earth with a thread of citric acidity. It reminded me of eating vegetables straight from my mother's garden throughout childhood—rustic, fresh, dusted with soil. Earlier I had managed to sneak a glass of this Gamay to Chet, the line cook, who would be dishing up eggs for at least another hour. It was part of our flirtation—I risked my job for his winks.

Exhausted from the busy morning, I was content to sit alone at the bar, sipping pét-nat. This wine style had become my new obsession ever since I'd started working at Reynard. The soft, fruity character of the "Pièges à Filles" was contrasted by its carbonated punchiness. Within minutes, I'd downed half a glass. It didn't have that sappy aftertaste I'd found in cheap sparkling wine. It was elegant, pretty, and aromatic, and it looked fantastic, too, with its cloudy haze.

A few weeks after starting the job at Reynard, I had taken a staff wine class with the wine director. With authority and calm enthusiasm, Lee explained how pét-nat was made through one single fermentation, which finished in bottle, resulting in trapped carbonation, as opposed to the Champagne method, which involved two fermentations. Furthermore, Lee had emphasized that these natural wines, whether still or sparkling, were unfiltered. Something called lees, a byproduct of fermentation, were responsible for that cloudiness. To me, the lees and the lack of filtration added a comforting texture—the wine felt nourishing, almost enough to quell my grumbling stomach.

I didn't want to waste precious wages on ordering food. Our staff meal, many hours earlier, had been tasty and robust—unlike previous restaurants where I'd worked in my early twenties, Reynard actually believed in providing nutrition to its workers. I was fairly happy in this job. It allowed me plenty of time to work on the novel I'd begun and attend fiction writing workshops in Manhattan. I also had a job as a nanny four afternoons per week. I wrote mostly late at night. My social life, then, was limited to my shift drinks after work.

"Hey, can I sit here?" To my surprise, there was Chet. He'd changed out of his whites and was now wearing an old IZOD Lacoste collared T-shirt and baggy jeans with sneakers.

"You're off early," I replied.

Chet shrugged and pulled himself onto a stool. There was that heat rising in my cheeks. Why did Chet's presence instill excitement in me? He'd never attended university, flipped eggs and made salads for about $12.50 per hour, and never said anything particularly romantic to me beyond, maybe, "You look nice today." Plus, he'd mentioned that he was very much married, although his wife lived back in North Carolina, where he'd recently moved from. And yet I felt very drawn to Chet's aloofness, his boyish good looks—ginger-tinted hair, green eyes, wry smile.

We made awkward small talk at the bar, complaining about various moments of that day's brunch—the many requests there had been for plain scrambled eggs, a dish that was somehow not on the menu—while I drank another glass of the same "Pièges à Filles," loosening up with each sip. Chet drank beer.

"You doing anything later?" The question came out of me impulsively. I saw Jarrett the bartender raise an eyebrow. I didn't mind him eavesdropping on our vaguely clandestine, budding romance—what did I have to hide? Chet sipped his drink, pondering my question. Had I been too forward?

"I mean, I'm sure you're working," I added hastily.

"Actually, I'm free." His tone was nonchalant. So was the way he then took out his phone and handed it to me to punch in my number.

As I was finishing, a hand rested on my shoulder. "Hooooo, and what are we drinking today," a voice boomed. It was Trevor—a waiter considered very senior at Reynard, as he'd helped open the restaurant and had previously worked for one year in some natural wine bar in France. He was on a first-name basis with some of the winemakers on our list.

Trevor's body language indicated that he was talking specifically to me, not Chet. Like most, if not all, restaurants, there were social divisions at Reynard—front-of-house didn't mix easily with the kitchen. Chet took the hint and began chatting with the sous chef, who'd sidled up near him at the bar.

"I'm having my usual," I told Trevor, gesturing to the glass. "Les Capriades."

"They really are masters of the form," said Trevor dramatically, pushing his glasses up on the bridge of his nose. "They age their wines for months before disgorging them, which makes them really elegant. Well, I was thinking about ordering a bottle of something nice. Care to join me?"

I agreed, and Trevor sat down. We were joined by a few other servers, and perused the wine list together.

"Oooh, let's do Matassa," someone said. It was decided that the perfect wine for this afternoon in midautumn would be something from a South African man named Tom Lubbe, who was making wine from old bush-trained vineyards in the Roussillon, a warm region in Southern France, with his label Matassa. As Jarrett opened a fresh bottle, I glimpsed the back label—this was a Louis/Dressner import, indicating that it was a somewhat well-established winery. Slowly, I was getting to know each importer's portfolio.

The wine was poured out—it was the color of a radiant sunset. This was a "skin-contact," or "orange," wine—made with white grapes macerated extensively on their skins before pressing. I held the glass to my nose and inhaled: kumquats, fresh and stewed peaches, all mingled in my nostrils. We clinked glasses, ordered some fries to share, and drank gleefully. My earlier days in New York had been trying for me as I'd worked my way through graduate school, juggled various part-time jobs while learning to do journalism, and battled loneliness and anxiety. I hadn't left all of that baggage behind me. But I had found an interesting community at Reynard, and now, as I savored this skin-contact French wine, you couldn't find a more carefree twenty-eight-year-old woman.

Back in my room in Bed-Stuy, freshly showered, I dressed for the evening. Chet had suggested a dive bar on Grand Street, where posh North Williamsburg met the seedier South Williamsburg. I threw on jeans, a ratty turtleneck sweater, and an old vintage leather jacket. I'd mostly sobered up from the afternoon drinking, but since it was chilly out, I opted to spend five dollars round trip on a metro ride rather than ride a bicycle.

I cast a glance at my laptop, sitting forlornly on the desk, which faced out the window to a new wine shop that had opened on our block. Bed-Vyne Wine was stirring up excitement, as it was one of the first places in the neighborhood where you could actually walk in and discuss your bottle selection with someone, rather than shoving money into the opening of a bulletproof glass covering and choosing between a Chilean Carménère and an Argentine Malbec. The fact that such a groundbreaking wine shop existed literally outside my bedroom door felt like more than an accident. It seemed to me that suddenly, a drink I'd once hardly noticed beyond selecting a boxed Merlot at Trader Joe's, was now a subject of great interest wherever I went.

It was dusk, and I could see people milling around in the shop, readying themselves for, I imagined, a cozy evening at home, perhaps cooking for friends or ordering in pizza while watching a movie. Through the glass, I spied Clara, the shop manager, talking excitedly to a customer about a bottle of wine. We'd chatted once or twice when I'd gone in to buy some Italian reds off the "$15 and under" shelf. She had mentioned that she was starting a weekly wine-tasting group. I envied Clara's position—to be knowledgeable, authoritative, about an entire array of wine bottles. To be able to sell them, confidently.

I headed out for the half-mile walk to the subway.

Chet was sipping Scotch at the bar. I ordered a Maker's Mark, neat. All around us, people in torn denim and puffy jackets, tattooed and pierced, drank beer or whiskey while awaiting their turn at the pool table.

Over the years, my life had become a complete question mark. I moved to New York to pursue graduate studies in cultural anthropology and had done well in the master's program, but not quite well enough to get into a funded PhD program. I wound up demoralized and in serious amounts of debt. Writing had become a passion for me, but living off it had proved nearly impossible. I had been working on a novel, my first attempt was poorly thought out, and I now tried to explain its plot to Chet. He nodded, seeming interested. He mentioned a book he was reading, and I sighed with relief. We could talk about books! From that point on, the booze flowed freely. At one point he excused himself and went to the bathroom.

When he came back sniffing, I could hardly restrain myself. "I would like some," I said without hesitation, smiling raptly. Chet returned the grin and slipped the baggie into the palm of my hand.

In the bathroom, which was adorned with chipped black wall paint and foggy mirrors, I used my house key to ingest small bumps and immediately felt the rush. I wasn't a cokehead; I'd only tried it a few times in my life. But that night I wanted to be bad. I wanted to punish myself for my failures—for screwing up my career and my finances. I wanted Chet, a married, relatively unpromising man.

Many puffs of white clouds and slugs of whiskey later, Chet and I went back to his place and managed some sort of unremarkable sexual encounter. Then we slept. At 8:30 a.m. the next day, I walked through the streets of Williamsburg and dragged myself into Reynard. I clocked in and began setting the dining room for brunch, wearing the same clothes I'd been out in, reeking of cigarette smoke and booze, and pretending not to notice the managers staring at me disapprovingly. Feeling sedated by the previous night's acts of self-disrespect, I served bowls of granola and plates of eggs Benedict, ferried ramekins of ketchup, and poured glasses of my newly beloved pét-nat to the wealthy and successful class of society, which I was sure I'd never join.

Within weeks, for that display of slovenliness and other minor but repeated offenses, management asked me to take my leave of their precious establishment.

When my job at Reynard ended, I sank into a deep depression related to my unemployed state, general lack of direction, and financial strife. After a few months and many therapy sessions, I managed to print out my résumé and hop on my bike, pedaling again toward Williamsburg. But this time, I was headed to a wine shop just down the street from

Reynard called Uva Wines. In the window, there were rows of empty bottles, placed as boasts: we drank this culty wine! For several years now, this shop located only two blocks from the busy Bedford Avenue subway stop had been one of Brooklyn's only sources for natural wine. I was determined to keep going on this path, even if I didn't know where it would lead me. I was hired almost immediately, and although I'd requested part-time work so I could continue to write my novel, I was started on six days a week.

As spring met the early days of summer, I quickly became used to riding my bike up to Williamsburg, then the demanding hours of standing in the shop, restocking bottles at the start and end of shifts, fending off questions from customers who knew more than me. But I was determined to learn, and whenever there was downtime, I researched every detail of every wine, read importer websites, and asked questions of my colleagues. I wouldn't let myself end this job in disgrace, as I had at Reynard.

Every Friday, it was the same routine: from 6 p.m. onward, when the first rush came into the shop—the after-work crowd—we were slammed. These were the creative professionals who had been living in this neighborhood for years—before it had become this circus of fashion models, visiting Europeans, and hipsters who lived off trust funds, and before upscale places like Reynard had opened down the street. These people, with their sophisticated palates, had long patronized Uva and loved to discover the natural wines we had in stock.

I ran around the shop excitedly, catering to each of these well-spending customers individually. If they were up for something new, I educated them on Partida Creus, the newly hot natural winemaking duo from Catalunya, brought in by upstart importer Alvaro de la Viña. Their bottles were graced cryptically with only block-letter initials. "The letters reference the heritage grape varieties the winemakers, who are an Italian couple with a background in architecture, rescue from abandoned vineyards. This one is made from Sumoll, a red grape, which was primarily grown for big, corporate Cava makers," I explained, selling a bottle nearly every time I told the story. I loved telling customers about how natural winemakers were iconoclasts, rebels, and rule breakers.

We also had two refrigerators full of expensive bottles, ranging from top-notch artisan Burgundy to Grower Champagne to aged Barolo. Once the first rush had passed and the store manager had left for the night, we went to the fridges to discuss what our splurge bottle would be for the evening. Buying the wine at its imported price, and being able to "taste" it on the job, was definitely one of the perks of working at Uva.

J.P., a guitarist who had fallen in love with wine while touring in Europe, made a case for his bottle: "2010, now that was a stellar vintage for Burgundy, and it should be just coming into the window of drinkability." It was a weekend night in June, and a streak of coolness in the air was inspiring thirst for something really special. J.P. ran one hand over his gel-slicked hair, getting excited about the prospects of a high-class Pinot Noir from a particular estate in Vosne-Romanée. He read about all of these elite, expensive wines on a website called Burghound, and expounded upon their graces to the rest of the Uva staff during each shift as he swirled a glass.

I eyed the classical French lettering on the label. It meant almost nothing to me, but I was definitely curious.

"Yeah, mate, that's a good one, but have you ever dug into one of these?" Charlie, a blond party boy from the English countryside, was fondling a wine whose label featured only a painting of a woman, and the words "Le Pergole Torte," along with the vintage, 2004. This was an Italian "unicorn wine" whose cost would have been in the hundreds of dollars. But we didn't really mind the price—we were consumed by thirst for these highly coveted bottles, to which we nobodies improbably had access.

I shifted my weight from one foot to the other, thinking carefully about what to say to my two colleagues, both of whom had wine knowledge above and beyond mine, although I was quickly catching up. "I'm in the mood for white—what about this?" I'd pointed to a bottle of Didier Dagueneau's cult Sauvignon Blanc, which by reading about I knew to be rare and a benchmark of high-end, collectable French wine.

My choice prevailed, and the Dagueneau was opened. J.P. poured it into a decanter, then filled our glasses. We swirled and hovered our noses over the moving liquid—the wine was powerfully aromatic. It reminded me of a dank cave. A dusty cave, somewhat rotten mold on its walls. The scent was not right. When I looked at J.P. and Charlie, I could see they were not enjoying the wine, either.

"It's corked," said Charlie flatly. My shoulders sunk with disappointment. We each poured our glass, laden with the taint known as TCA, which happens randomly and indiscriminately to wine corks at an alarming rate of up to 10 percent, back into the bottle. Our collective mood darkened momentarily.

J.P. went to the fridge, not the fancy one but the one where we kept everyday-drinking wines, rummaged around, and popped open a very different Sauvignon Blanc—this one was from the Loire Valley *vigneron* Thierry Puzelat, whose wines I knew from my tenure at the restaurant. My colleagues seemed unimpressed as they sculled the savory white wine, at least a hundred dollars cheaper than the first one we'd opened (which, being corked, we could return to the distributor without expense), but I was actually thrilled. It was partly that I'd avoided spending my entire night's earnings on a fancy bottle, but the fact was, despite my growing affection for expensive cult wines, simple and humble natural wine was what I truly loved. Whenever Uva's owner brought us out and poured twenty-year-old Burgundy and Barolo for us to blind taste, I gave my best guess at the year and the producer, trying not to seem as clueless as I felt. But ultimately, a basic French *glou-glou*, with its searing acidity and wild, foresty characteristics that reminded me of the camping trips of my childhood, was more my style.

I finished my first glass and felt it energizing me as I approached a customer, who was browsing the fridge. "What kind of wine are you looking for tonight?" With any luck, I'd upsell him from his go-to cheap bottle to something more singular, made from organic grapes with the faintest touch of a winemaker's hand, and potentially, forever change his palate.

As summer neared its end, I had finished my novel manuscript and was going strong at Uva, working five days a week. One night, out drinking with the shop owner, I'd accurately blind-guessed the vintage of a Chablis as 2009. I had been diligently tasting and reading while at work, and on my nights off, I attended biweekly tasting groups hosted by Clara, who ran the wine shop across from my bedroom, at her home not far from mine in Bed-Stuy.

I'd become one of the best salespeople at Uva. I knew the *crus* (appellation-classified hills) of Burgundy and Beaujolais by heart and could even recite the biographies of a few esteemed producers. I could effortlessly guide the rare wine connoisseurs over to the fridge to select an aged Nebbiolo or vintage Grower Champagne.

Naturally, since I was no longer desperate for intellectual stimulation, I'd found a new distraction: J.P. We'd become close over the months, as our shifts frequently overlapped, sharing jokes about annoying customers and engaging in long debates about the meaning of "terroir." Toward the end of August, it was announced that J.P. was going to be promoted to manager with buying privileges, and he was in a good mood about this, although he questioned to me in private whether it spelled the end of his music career.

At least once a week, J.P. and I went out drinking together after work. Occasionally, over our late-night glasses of Gamay and Pinot, he would spill his guts to me about troubles at home. He was in a long-term, live-in relationship, and earlier that summer, J.P. had met a cute French girl and they'd slept together several times. It had enthralled him and made him feel terrible at once.

"So, we've decided to break up," J.P. confessed to me one night post-shift, at August's end. I listened, swirling my glass, which contained Nero d'Avola from Vino di Anna, a Sicilian winery run by an Australian woman (Anna) who reportedly pressed the wine in a traditional concrete structure called a *palmento*. We had brought the bottle into a local bar, where they let us BYO as long as we poured some for the bartenders. The wine tasted like ripe strawberries laced with fragrant, burning incense.

"I just wish I could be with someone like you," continued J.P., "who I can talk about wine with." He blurted it out in a way that simultaneously seemed impulsive yet very targeted and intentional. I nodded and drank my wine as if I'd hardly noticed the comment.

I knew I should be wary. This job felt like the only thing I had going, and sleeping with the manager could lead to problems. And yet I was so attracted to him—for his wine knowledge, his height, and the way he placed an arm just briefly around my waist and squeezed lightly, whenever I remarked intelligently on wine during our tastings at the shop. Being young and lonely, I blocked out any voice within that may have informed me that his attitude and actions, as my colleague, were questionable at best.

"I have an idea," he said. "Why don't we go back to the shop and get one more bottle, then go meet Charlie? He's DJing a party for some people who work for David Chang, and I'm sure he can get us in."

As we walked the four blocks toward Uva, the warm breeze felt comforting on my bare skin—I'd changed after work into a loose spaghetti-strap dress. I noticed J.P., who was a full foot taller than me, glancing down at my body, especially my chest. I thought about what he had just told me, about his relationship ending. In the Brooklyn summertime, rational thinking could easily be buried under the sway of tantalizing desire.

J.P. unlocked the security gate and raised it halfway so we could step inside the shop. We turned on just one light so we could maneuver down the creaky, old stairway to the cellar.

The cellar at Uva was a treasure trove. There were dust-laden bottles from the '70s hidden amongst the stacks, some forgotten entirely. We began poking around in the open cases of wine, which sat haphazardly on shelves, sharing our findings from different parts of the cellar:

"Are you in the mood for Beaujolais?"

"Maybe. Have you tried anything red from Austria?"

"Love a good Zweigelt. You know, I wouldn't mind a white Burgundy."

"Ooooh, that does sound good. But when was the last time we drank Chenin?"

Then, we stopped—our meanderings had brought us face to face. I shivered and felt goosebumps rise on my arms. J.P. leaned down and kissed me, and I returned the affection, reaching my arms up around his neck. He paused and mumbled something, a kind of apology. Probably, "We shouldn't, but . . . " But of course, we did.

J.P. ran his hands up and down my body, then without taking his lips off mine, he lifted me up and placed me back down on a stack of boxes. They swayed under me, but he held me firmly. Soon, his pants were at his ankles and my underwear was on the floor. After all those weeks of flirting, it felt like the sweetest release. Every bottle of Burgundy we'd consumed together had been an aphrodisiac—the floral notes of Pinot Noir had warmed my blood, and J.P.'s impressive knowledge had charged my groin with desire. Apparently, he'd felt the same way.

He'd flipped me onto my stomach, and was struggling to make it work, when suddenly we heard a massive crash a few steps away. We stopped and gasped—three or four cases of wine had toppled, and there was broken glass and wine everywhere. It was pink, upon closer inspection.

"Oh, thank god," sighed J.P. "It's only the Château Peyrassol." Mid-priced rosé. We laughed, cleaned up the mess, and got ourselves out of that cellar as quickly as possible.

Soon, my shifts at Uva became unbearable. I was sure that J.P. had leaked the story of our intimacy to our colleagues—even the stock boy smirked broadly one night when J.P. was pouring me a glass of Chablis. As J.P. assumed his role as manager and became a decision maker in the shop, our conversations at work became stilted and awkward. I thought of our affair as something I'd gone into willingly and only held the faintest notion that there was any wrongness, including on his part as my now-superior, in what had happened.

I began to focus on figuring out next steps in my new trajectory of a career in wine. At some point, I would need to work someplace other than Uva. The hourly rate wasn't enough despite a fifty-cent raise the previous manager had granted me, the work was intense, and J.P. was getting on my nerves. I'd been writing for a website called *Food Republic* with moderate success, profiling winemakers and offering 101-level explainers about wine regions or types. It paid almost nothing but was incredibly satisfying.

One morning in early September, as I sat at my desk reviewing my novel, wondering whether it had enough merit to bother sending around to agents, an email came: my editor at *Food Republic* wanted to know if I was able to accept a press trip in the next few weeks, to Burgundy. Was I available? I wrote back more quickly than I'd ever replied to an email. Then I rode my bike to Williamsburg, flying along the streets, elated. I was going back to France! I'd only been once, as a university student, visiting a friend who was doing a year abroad in Paris—and I'd held onto a strong desire to return. Now I was going as a journalist—and it was paid for. A dream!

By 6 p.m., the entire journey had been confirmed. I could not contain myself.

I bragged to J.P. and anyone who would listen about the upcoming trip. Then I realized I needed to ask for time off. It was granted to me with a slight eye roll.

That trip to Burgundy, followed by one week in Paris, gave me a taste of the wine writing life and ensured that I would do anything to make it work for me.

We visited prominent estates whose moldy bottles of greatly expensive wine had been stacked in cellars for many generations. But it was during a few days in Paris that my love for the country and its natural wine scene was confirmed. Arriving on a train from Chablis, I crashed on an acquaintance's couch and proceeded to get violently ill from the vast quantities of foie gras and raw-milk cheese I'd consumed over the past week. After downing many bottles of the French sparkling water Badoit, I managed to step out into the sunny September weather and explore the Marais, with its overpriced fashion boutiques and charming cafés.

Paris seemed more welcoming than it had when I was a twenty-year-old college student with baby fat clinging to her hips and a closet full of blown-glass bongs. Back then, I'd hardly managed to pronounce *bonjour*. The locals were not impressed. In New York, as an anthropology graduate student, I had finally taken courses in French, as part of my research on Senegalese postcolonialism. Now, thanks to my nascent linguistic skills, Paris showed a friendlier face.

Before leaving Paris, I rallied my stomach for one proper meal at a place called Saturne, which was reputed to have an "all-natural wine" list. The tasting menu was a whirlwind of flavors and textures, and the wines, ranging from Jean-François Ganevat's estate Chardonnay to Frank Cornelissen's volatile Mount Etna red, were beautiful enough that I almost forgot the aristocratic notes of Pinot Noir I'd encountered in Burgundy. These were not glou-glou wines; they were richly textured, complexly flavored, entirely naturally made wines.

On the plane back to New York, I revisited all the moments of gustatory ecstasy of the past fourteen days. The cellars of Burgundy had been wonderful, but it was the grit and utter coolness of Paris and its natural wine scene where I felt more at home. I knew then that I would continue to write about artisanal and conventional wines, as needed for my career development. But natural wine, made with only grapes, had taken hold of my heart and soul. I wondered when I would have another chance to return to Paris, where every neighborhood on the bohemian Right Bank had its own collection of gorgeously outfitted natural wine bars, shops, and restaurants. And I swore to myself that one day, I would figure out how to get back to France, to live and write.

— Two —

A Way Out

2017

A wintry March morning wasn't the ideal time to take a leisurely bike ride through Brooklyn, but the chilly weather was triggering my coupling instinct. A surge of hopefulness fueled me as well, coaxing a naive optimism to overrule my own sensibilities and the reality of past experiences. I would brave the winds.

Before leaving the apartment, I bundled up in a coat and scarf. Then, impulsively, I undid the scarf and ducked into the bathroom, where I smudged brightening cream under my eyes, put on blush and mascara, and applied a shimmery lip gloss that belonged to my roommate. I blinked in the mirror at this slightly improved version of myself.

Out on the street, the expressway roared above me. I mounted my ancient cruiser and rode past my laundromat, a bakery, and the coffee shop where I liked to write. Briefly, I considered a looming deadline that would bring me to the coffee shop later that day—a monthly column I'd been writing for the national food website *Eater* that looked at a different wine variety each installment. But as I turned down

Metropolitan Avenue toward Evan's apartment, my thoughts focused intensely on him.

Several weeks had passed since our encounter in France. My text messages had gone unanswered, and I hesitated to send more for fear of seeming desperate. A casual visit seemed like the best option. "Hey Evan, I was just at a café nearby and thought of you," I pictured myself saying. He'd invite me in for a coffee. We'd remember the passion of that late night. It would be rekindled. Maybe later, we'd head out for a bottle of wine to bring back to his place, where we'd cook dinner.

Pedaling along the dingy pavement, I relived the thrill of that wine-soaked evening back in Angers. I'd felt high on life, having made it to the famous "salons de Loire," the annual tastings in France's vineyard-dotted Northwest where the best natural winemakers in the world showed their recent vintages.

Of the multiple natural wine–focused "salons" (tastings where the growers pour their own wines) taking place over the first weekend in February, the most famous and iconic is la Dive Bouteille, affectionately known as "la Dive." I'd wanted to attend la Dive ever since hearing it mentioned in reverent tones at Uva. As my writing career had grown, I'd traveled several times in Europe, the US, and even South Africa via standard press trips for general wine writers, which were always sponsored by official regional boards and showcased a wide range of producers. That sponsorship meant they were all expenses paid but also that they generally didn't allow visits to winemakers in the "natural" realm, as these individuals and families nearly always worked outside of appellation systems, subsisting on an alternate market.

Finally, in part because of this *Eater* column I'd been writing, a casual acquaintance named Phil had reached out. Did I want to join him and another sales associate from Jenny & François, the natural wine importing company, on their annual trip to la Dive and the other salons? It was a thrilling offer. J&F was well connected in the Loire

Valley, which, I knew from previous travels, was home to a rustic and eclectic culture of natural winemaking. For nearly a decade, Jenny Lefcourt of J&F been importing wines by radical, anti-establishment natural winemakers, individuals and families who refused to conform to appellation standards such as adding yeasts, chaptalizing—meaning dousing a wine with sugar so it would have more booze—or otherwise altering the process.

These producers did not have tasting rooms where young employees served cheese plates for road-tripping tourists; they did not make wine in châteaux of the bygone aristocratic era. They lived rather simply, working with their hands, and honoring (or, in tough years when hail hit the vineyards, suffering from) the whims of nature as well as the traditions of previous generations. They made wine as their grandparents did, before the vineyard owners in France began to pump their soils full of chemicals, eager to increase production (and encouraged by the government to do so) after the devastations of World Wars I and II.

"Will you be at la Dive? I'm going with the J&F crew," I had said to Evan at our weekly French class on Tuesday nights. His response was enthusiastic. Of course, he would be at the salons. He'd see me there. Couldn't wait.

It's difficult to explain, if you're not already obsessed with natural wine, why it's so exciting to spend hours on your feet in a poorly lit, unheated, limestone cave dug into a hill, wearing a full coat and hat and jostling with attractive, slightly standoffish Europeans with your glass outstretched.

But if you've caught the bug and desire nothing more than to meet the makers of these potions, to try speaking your best French so you can finally ask when they started using amphora, or if they'll ever bottle that single-vineyard Pinot again, then these two or three days of epic, exhausting tastings will be the highlight of your winter, or maybe your year.

Now over twenty years into its existence, la Dive has evolved from a gathering of a few dozen scarcely known producers in the town of Saumur, in the heart of Cabernet Franc country, to an international, multiple-day affair attended by importers and sommeliers from as far as Moscow and Tokyo and California. It has grown so much that not only are there French winemakers pouring in those romantically lit caves, there are also Milan Nestarec from the Czech Republic; Arianna Occhipinti from Sicily; and Le Coste, north of Rome. There are even "off" tastings held at alternative locations, as la Dive itself is unable to accommodate the new generations of natural winemakers.

One of the salon weekend's most treasured aspects was the opportunity to party in the company of international natural wine celebrities. La Dive Bouteille felt like the Met Gala, except there were no actual invites, the costume theme was clearly "hobo," and the only paparazzi were Americans like me snapping photos of every cult bottle we drank and Instagramming them shamelessly.

At the end of each tasting day, the sommeliers, importers, aspiring winemakers, and journalists like me had no choice but to continue quenching our thirst, with more and more wine. My friends and I were exhausted, but we knew that if we went back to our hotel to sleep, we'd be missing out. Around 9 p.m., we found ourselves in the small city of Angers, on a particular street where two natural wine establishments faced each other. In the car-free zone between them, people stood in small circles, greeting each other with kisses and pouring from magnums, not minding the cool air.

Phil, who had been to the salons several times, was smart enough to book us a table at the bistro Chez Remy, famous for its deep cellar. It was gustatory hedonism at its finest, and though we knew we'd have heartburn (and a hangover) the next morning, we couldn't hold back. Around midnight, we were on to the cheese course and giddy in that unique way wine can make you feel, when the man I was hoping to see walked in: Evan. It was easy to spot him—he was tall with a head of curly hair and puppy-dog eyes, and a smile that could make me weak in the ankles.

Evan and I had been flirting for months, all through the past autumn, during the French classes we both attended on weeknights in Brooklyn. We'd known each other for some time. He worked in Manhattan, at a wine store owned by his father, which had been an early champion of natural winemakers. Back in autumn, following a large public natural wine tasting, Evan threw a big party at his apartment where magnums of Beaujolais had been poured liberally. After that event, he let the word out that he found me attractive. It felt very promising.

As the nutty, rich Savagnin in my glass coursed through me, I watched Evan make his way around the bistro, greeting all the winemakers he knew, and I felt emboldened. This was the moment to make my move.

"Oh, hey, you!" I gave my most coquettish smile.

"Heyyyyyy." His voice was low. Tired, sexy. Eyes scanned me up and down, a hand clasped my upper arm. We kissed on both cheeks.

I asked him what the day's highlights were, and he babbled about wine for a few minutes. But I didn't wait long—I went in for a kiss, and the connection was devastatingly hot. Within moments I settled up my bill and said good night to my dining companions and was walking back to my hotel with Evan.

Our intimacy that night felt exciting and strange all at once—there was something missing the whole time, but I couldn't pinpoint it. I'd never really known trusting, committed love before, so I couldn't be sure how to identify its absence. And maybe I was, to some extent, in denial—I didn't want to see how much our interaction lacked real grounding. Evan did not stay the night, and I woke up remembering our affection almost as a dream.

All the next day, as I roamed the caves at la Dive on its second day, passing from barrel to barrel and seeking out the most exciting producers to taste with, I was filled with that tingly sensation of newly consummated attraction.

I felt that excitement again back in Brooklyn as my bike turned the corner, past the hair salon that served single-origin coffee to its clients, catching a brief glimpse of the wine bar, the Four Horsemen—I'd had to ban myself from that establishment after the previous tax year, when I went through receipts and realized how many hundreds of dollars I had spent on bottles within. I stopped pedaling in front of Evan's loft. Why couldn't he even bother to offer me a simple reply? Didn't I deserve that? Perhaps he was sick. It was New York—you never knew what was going on in a person's life.

I ran my fingers through my hair to relieve the helmet's influence, nervous as if I were back in high school and I'd decided to ask my crush to prom. Evan and I had real chemistry, didn't we? We'd flirted so much during French classes. Our connection that late night in France had been passionate, as I remembered it.

It had been about three years since I'd had a steady relationship. I really wanted this to become something.

I knocked. A woman opened the door, blinking as if I'd awoken her—I recognized her as Evan's roommate from the party.

"Hi, I was just nearby and thought I'd see if Evan was around." I was trying to play it cool. The woman's face scrunched up in confusion. I felt my heart twist, preemptively.

"Oh!" She squinted at me with vague recognition. "Well, he doesn't really stay here much anymore. He basically lives with his girlfriend now. Do you . . . need his cell number?"

I stammered some kind of a reply and quickly buckled on my helmet with shaking hands. The word "girlfriend" reverberated in my head. I wished desperately I was headed anyplace but to the shitty apartment I called home. How would I manage to sit down and concentrate on writing about wine? I felt pathetic, recognizing that this was a pattern I had fallen into—it was hardly the first time I had pursued someone unavailable to me. Hadn't I done the same thing with Chet? With J.P.? Why was I doing this to myself? Even if Evan had hidden the fact he was in a relationship, it was also true that he had not shown me genuine affection or any desire to be with me.

For whatever reason, perhaps simply because I needed *something* to blame, New York itself seemed at fault. Every time I tried to stand on my own in that city, to be comfortable and confident, it threw me another demeaning or low-paying job, another shitty yet overpriced apartment, and now, yet another heartbreak.

Work at the moment was not exactly booming with success: for months I had been waging an unnerving email battle with a large media company that refused to acknowledge the fact that they owed me for six months' worth of writing. The total amount, around $2,300, surely wasn't much to them. But I had hardly that equivalent in the bank. Rent was due. I wanted new shoes. I needed money to go out, because if I stayed in that cramped apartment and watched Netflix and cooked spaghetti for too many consecutive nights on end, I would get depressed.

Every week for the past year, I had cycled over to Bushwick and entered the home office of a young psychoanalyst of Lacanian persuasion who charged me fifty dollars to lie on his couch and work my way through childhood pain, generally culminating in one sentiment: "I've gotta get out of this place." I was over New York. But I had no idea how to make the move.

As winter gave way to spring, instead of falling away, my sadness morphed into fury. It was directed partly at myself, partly at the devastating results of the 2016 presidential election, and primarily at New York, this city I'd naively trusted when I'd moved there at the age of twenty-four, and which had continued to betray me. Its men walked all over my heart.

Its media industry regarded me as an outcast freelancer, leaving me to struggle without income from one article to the next. Although they gave me assignments from time to time, I felt that none of the established wine magazines took me seriously, because I wrote about natural wine. And natural wine wasn't normal or legitimate to them, from what I could tell. It was niche, marginal, for the freaks who cared about things being grown organically. Real wine was about status and wealth—that's what the old guard, who controlled the media, stood for.

Which all conspired to have me feeling rather shitty upon learning that my crush, whom I'd impulsively slept with, had a girlfriend. I fired off a few text messages to Evan letting him know that I was onto his betrayal. His eventual reply was evasive and uncaring: "well, i said it had been awhile since i'd been with someone but i didn't say i was single. i am sorry if u were hurt." Whatever it meant, this absurd statement that he was apparently too cool to even punctuate properly, it affirmed what I already felt, had been feeling for months: it was really time to get out of New York, and out of the US entirely.

The clock was ticking: as my twenties clambered on toward my third decade, I'd increasingly wondered when I would find a committed partner. I wanted someone to share life's joys and tribulations—and if I was honest, to share the rent and more practical aspects of living. And I wanted a family of my own—maybe just one child, so I could remain somewhat free to travel. Somehow, the City's vastness, its never-ending downward spirals of online dating, its plentiful supply of attractive and smart women, defied my need for coupledom.

"He keeps saying that my career doesn't matter, because I'm only a bartender," said Gaba, looking down into her drink, twirling the little straw.

I'd called her to complain about Evan, but it turned out that she wasn't having it any easier. We had met over Manhattans near my apartment—mine up with a cherry, hers up with an orange slice.

I'd known Gaba for years through various common friends and also through the restaurant industry—she was an expert mixologist and wine lover, currently working at June, Brooklyn's latest spot for guzzling glou-glou.

"So, he's a hacker, and that's a better career?" I replied.

Gaba shrugged. "He makes three times more money than me."

And yet this boyfriend of hers refused to cover many of their shared living expenses. As we ordered a second round, I realized that Gaba needed comforting far more than I did. Evan was a jerk, but I could brush him off—meanwhile, Gaba's live-in boyfriend had been driving her to the breaking point with his persistent judgmentalism.

"Why don't we go back to my place," I suggested as we downed our drinks. I could see tears were not far off, so we quickly walked around the corner to my apartment, where we sat on the rim of the bathtub, smoking with the window cracked open.

Gaba's cheeks were streaked with mascara. This was a talented, smart woman—how could she put up with such bullshit? Then again, I had to ask the same thing of myself. I could only fixate on one thing—we had to leave New York.

"Look, you have something going for you that he doesn't," I told Gaba. We exhaled out the open window, toward the roaring freeway, and drank the glasses I'd poured from an Oregon Pinot I had open on the counter. "You have a French passport. Are you kidding me? Why are you still here?"

A few years ago, Gaba had helped a French colleague who needed US citizenship, and the result was that she could legally live and work in France, something unavailable to most Americans. Gaba swabbed her eyes. I could tell she was embarrassed by the display of emotion and that a man could destroy her otherwise strong countenance. This was the first time we'd had a moment together like this—I'd generally seen her in professional contexts, either making and serving fancy drinks or enjoying them. Gaba and I had once or twice been in France at the same time and I'd noted that she spoke the language well and always dressed impeccably.

"If I were you, I'd move to Paris right now and get a job," I said. "You can meet me there. We'll share an apartment."

"All right," said Gaba, meekly at first, then raising her voice. "Let's do it—I've been in New York for too long anyway." We tossed our cigarette stubs into the toilet and swore that we would dig our way out of the City's sinkhole, rather than falling deeper in.

France, I knew, was not paradise. There had been horrible, scarring attacks on French culture in recent years—the *Charlie Hebdo* shootings, the Bataclan suicide bombing and shooting, and others. Marine Le Pen was on the rise. But as I lived by the grape, France was my spiritual home. Paris also made me think of literary greats—of Ernest Hemingway, of Mavis Gallant, of Gertrude Stein, of James Baldwin. It spoke to me of intellectual freedom, as it had to North Americans of past generations. I now knew enough French to get by, and true fluency was within reach.

I began to work on something called a visitor's visa, which would entitle me to a year of living in France, but it required me to have an apartment already secured in that country, something virtually impossible without already being there. So, I would live illegally in Europe, then. At least until I could get the visa. As my sense of ennui deepened, I no longer cared about rules or common sense. I would do anything it took to start moving in a new direction.

I gave my roommates notice and began selling off my meager possessions—the old cruiser, my IKEA bedframe and desk. A few things went out onto the curb, below that highway. My beloved books from graduate school and my subsequent year of studying fiction were packed into boxes, and unable to part with them, I cajoled a friend into storing them at her apartment.

Gaba promised she would join me overseas in a few more months. On a warm day in early May, I rode my cruiser, which had not yet found a buyer, one last time across Brooklyn to June to see her. On the way, I wondered if I would miss these streets and the daily life that animated them: people chatting on stoops outside the borough's signature brownstone buildings, the industrial beauty of the Williamsburg Bridge and the Navy Yard, the coffee shops where baristas frowned at customers who tried to order "mochas" or lattes in the afternoon, the Trader Joe's where I stocked up on pasta sauce and bags of oranges. But I thought also about Paris and the little things I loved about that city—like the ornate typography on the façades of bakeries and pharmacies, as if they were the most important things in life—and felt that I could let go of all this, easily.

June was lightly buzzing with the after-work crowd as I sidled up to the marble-topped bar. Gaba came over smiling with chapped lips, her tiny hoop earrings shimmering in the candlelight. She poured me a glass of an Austrian natural wine made of Gruner Veltliner before I could order anything. The rich, lemony taste filled my mouth. We turned our conversation to our future lives in Paris.

"I think that the owner here might invest, if I opened some place abroad," Gaba murmured quietly. My eyes widened, picturing it: our American-run *bar à vins* making a splash on the scene in the 11th or 10th arrondissement, attracting all the ex-pats. Late nights of popping magnums and riding home on the back of motorcycles. Weekend trips to Champagne.

"That would be amazing," I said. "I'm sure I can get a work visa eventually and in the meantime, I can help you open it . . . as a consultant?" It seemed plausible enough. Gaba and I knew France somewhat, from our various trips there. Opening a wine bar together was a pipe dream, perhaps, but it was also something to guide us along in this impromptu move.

But before I could immerse myself in Gamay and baguettes, I had one more trip planned: an eight-day, natural wine–focused excursion in Georgia—the country, not the state. It was someplace I'd never been, and never would have thought about going, had it not been for an importer of Georgian natural wines inviting me along. The trip would include sommeliers, other importers, and winemakers from around the world, visiting only natural winemakers, in this Eurasian country that holds the world's oldest continuously producing wine culture. This opportunity invigorated me after such an emotional past year and its many disappointments.

With everything ready for my move—a sommelier friend finally bought the cruiser—I embarked on a long flight via Istanbul to Tbilisi. Afterward, I'd make my way to France, and though I didn't totally have my life there figured out, I was entirely confident it would be better than my tired New York existence. It would be new, it would be free, it would be drenched in vin nature.

– Three –

A Kiss in Qvevri Land

My cell phone alarm sounded, and I rustled an arm toward it, trying to remember where I was. I felt starchy sheets—a hotel bed. I slowly remembered the long flight, finding my way to my room, briefly brushing my teeth before crashing. Opening my eyes, I saw the room come into focus, noting the simple wooden frame of the window, the humble décor, and my suitcase open on the floor. The clothes I'd worn on the plane were crumpled in a pile. I sat up and glimpsed the view out the window: below the hotel was a labyrinth of winding streets curled around the red rooftops of low-rise buildings. Farther away, the hillside was dotted with trees, and I saw a castle in the distance. Tbilisi appeared like an archetypical medieval town, at least here in the old part of the city.

The second leg of my flight had been delayed in Istanbul, and so the time difference was aggravated by the stresses of long travel and only three hours of sleep. I desperately needed coffee.

In the breakfast room down the hall, I sat near one table of early risers already working through an enormous spread. A woman came over and began covering my table in plates: dishes of sautéed greens, fresh white cheese, scrambled eggs yellow as a canary, and slices of rustic, warm bread. I drank coffee and ate hungrily. Meanwhile, the table beside me was chatting earnestly, clearly far more rested—so, they had not come from the US then—and I struggled to identify the accents of the speakers. Intrigued, I turned to one of the men. He had a scruffy salt-and-pepper beard and wispy blond hair that stuck up on his head, like a baby bird's fluff.

Never one to let timidness overshadow curiosity, I leaned over.

"Excuse me," I said. "Where are you all from?"

The man with the unruly beard smiled semi-politely and flashed his very blue eyes at me. "We make wine in Australia."

"Which wines?"

Curtly, he said, "I make Lucy Margaux wines, and he's Tom Shobbrook."

"Oh, thanks." They resumed their conversation, and I went back to caffeinating myself. A rush of admiration came over me: I was in the presence of natural winemaking cult stars.

Twice before, I had tasted these enigmatic Lucy Margaux wines: once, a winemaker friend in Portland, Oregon, emerged from his kitchen with a bottle I'd never seen before, featuring a label made of fabric-like paper, with a hand-drawn, childlike illustration on it.

"This is so hard to find," he said, pouring out two glasses of the cloudy, pink liquid. We tried it, and I gasped.

"That's . . . the most natural wine I've ever tasted." I didn't know how else to describe it. The essential wine-ness of the wine, its pure form, came through in a way that I'd never encountered. It was entirely sulfur-free, which accounted for its vibrancy. Sulfites have the effect of killing off yeast and bacteria, which makes for a more stable wine.

Sipping more, we detected a streak of "volatile acidity"—generally considered a wine flaw, but I found it endearing. Like many natural wine drinkers, I was willing to forgive minor flaws because I so loved the liveliness of an unadulterated wine. Natural wine's unpredictability was part of its attraction, while conventionally made wine, "flawless" though it may be thanks to stabilization, filtration, and added preservatives, made me fully yawn.

That Lucy Margaux Pinot Gris had a spectacular energy that moved me. It was also refreshingly low in alcohol, unlike most Australian wines I knew of—we finished the bottle easily.

Some months later, at the tail end of a 2016 press trip in the Rhône Valley, I spent a few days in Paris, crashing with an American friend whose family had relocated there for work. One afternoon, I gathered with a few Parisian chef and wine industry friends at a beloved eatery called Les Arlots. The crew included Gaba, who coincidentally was also visiting France for that same Rhône Valley tasting. By the time our main course was due to come out, we had thrown back a bottle of pét-nat and something white. The conversation was flowing, and we were ready for more wine.

Just before the signature house-made sausage and mashed potatoes, drizzled with a special gravy, hit our table, the owner, Tristan, appeared. He held five bottles of red wine by their necks for us to choose from. Like many French bistros, Les Arlots had no wine list. Instead, you had to trust the person offering wine to you. You had to be willing to take a chance on a natural wine, in other words, without knowing every detail about it before it was uncorked. Tristan didn't even tell us the prices as he ran through them, briefly detailing where they were from, or a small note about the vineyard. He had many tables to attend to and not much time—we had to choose quickly.

A sommelier at our table immediately pointed at the only foreign bottle in the lineup. Slowly, I recognized the hand-drawn style of label—it was another wine from this Lucy Margaux, which I still knew nothing about. The bottle read only, "Vino Rosso." It was poured and we clinked glasses raucously. As I took a sip, the conversation around me grew muted, and I was drawn into the complex world evolving in my glass. The wine was stunning—it had flavors of fresh ripe fruits and a sour note of acidity that made me thirsty for another glass as soon as I'd finished one. We drank a lot of French natural wine that day, but the vineyards of Australia had captured my mind. I thought about that Lucy Margaux bottle for weeks—how the light acidity balanced with its rich texture and peculiar combination of savory flavors and fruity overtones.

Years later, on this cool Georgian morning in May, here I was in the maker's presence. Tom Shobbrook was another name I knew. He made notable Shiraz wines—as the Aussies call the Syrah grape—as well as some skin-contact whites I had tasted in New York.

The other participants on our trip stumbled groggily into the breakfast room to throw back cups of coffee, and soon we were called to our transport. I boarded one of two busses and found that the Aussie winemakers had gotten on the same one. Everyone mumbled greetings to each other in soft morning voices. Despite the gentleness, I suspected that this trip was going to be a wild one.

Every press trip I'd been on, I'd traveled largely with your standard wine journalist types: older, white males who prattled on endlessly about which Bordeaux vintage was better or whether the prices of certain Burgundy estates had finally peaked; there were also plenty of hobbyist bloggers who had other jobs to sustain them and used these junkets as free vacations, returning home to write essentially a sponsored post on whatever region we visited.

Here, now, were people who ran natural wine bars in Oakland; imported natural wine into Barcelona and Berlin and New York; or made natural wine in Oregon, Australia, and Italy. There were only two journalists on the trip: me and effectively the only person out there who had made a career writing about natural wine, the outspoken author Alice Feiring. We settled in for the two-hour ride to Imereti, the eastern part of Georgia.

"Our first stop today will be at Zaliko's—he's the *qvevri* maker," our host, John Wurdeman, announced. A few of us nodded knowingly—of course, as natural wine drinkers, we had heard of these ancient clay vessels, qvevri (pronounced "kvevri," with the first *v* said very softly). Although a handful of European winemakers had adopted qvevris' usage in recent decades, they were unique to Georgian winemaking culture, possibly dating as far back as 6000 BCE. The qvevri and their exoticness were a large part of why many of us had journeyed to Georgia—we had seen or used barrels, concrete tanks, and stainless-steel tanks to our heart's content, but clay was something different.

John stood at the front of the bus with dark bags under his eyes and his long hair pulled back into a ponytail. An imposing, charismatic American who had found his way to the western part of Georgia as a young fine arts painter, John now made wine under the label Pheasant's Tears. With one foot in the Western world and the other here in Georgia, John had become a sort of global ambassador, representing the country at wine fairs in London and Paris and bringing groups like ours to visit.

He tossed his hair over one shoulder and explained that Christianity held much sway in Georgia. "You need to know," he said, "that guests here are seen as 'a gift from God.'" We would be welcomed heartily with food and drink, and lots of it, at every place we stopped. "So, pace yourselves," he advised. There would be toasts, led by a "toast master" called a *tamada*, at each *supra*, as the Georgians named these feasts. And, John told us, as the last urban buildings gave way to green hillsides and rough, poorly paved roads, there would be the famous *cha-cha*, a strong home-distilled grape brandy that Georgians loved to drink with abandon. "Careful—that cha-cha is serious stuff!" Alice and a few others called out, "Thanks, John!" He smiled warmly and sat down.

The bus veered off onto a road that was desperately in need of new paving. As the scenery became increasingly rural, I turned around in my seat to survey my bus companions. There was an exotically beautiful, red-lipped Italian woman whose family made wine on an island off the coast of Tuscany. I already knew Alex, a tall, affable guy who did the wine list at one of my favorite Brooklyn bars. I heard Tom Shobbrook talking excitedly in his Aussie accent to Jill, a pixie-haired wine consultant from Minneapolis. In front, Alice was chatting quietly with John—the two knew each other well from Alice's many trips to Georgia. And sitting right behind me was the Lucy Margaux winemaker.

After that one bottle in Paris that I couldn't forget about for weeks, I'd researched Lucy Margaux online but hadn't found much. His nickname, I learned on social media, was "Wildman," thanks to his mad-scientist hairdo—which today was on fine display, with strands reaching out in all directions—as well as a tendency to stand atop bars and recite poetry on a whim. The name also referenced the unapologetically natural, indeed wild, wines he made in South Australia.

The Lucy Margaux website mentioned biodynamics—but I'd heard from asking around that the Wildman's vineyards apparently were not his own. He was a négociant, working directly with vineyard owners to harvest and purchase the grapes they grew, and therefore Lucy Margaux Wines did not bear any kind of organic or biodynamic certification. None of this seemed unusual. Many natural winemakers do not own the vineyards they farm or make wine from, and many also eschew certification, grumbling about the costs and the paperwork involved. But I was intrigued by the mention, on the winery website, of biodynamics—a fascinating approach to agriculture.

"Hey, do you mind if I ask you a question?" I blurted out over my shoulder.

Wildman was staring out the window with a forlorn expression, and he slowly shifted his gaze to me. "Sure," he agreed.

"I heard that you practice biodynamics in your vineyards, and I'm wondering what exactly that means to you, or how you approach it." It was a question I had enjoyed asking winemakers over the years, as everyone had a unique approach to this esoteric agricultural philosophy founded in the 1920s by Rudolf Steiner, an Austrian polymath and spiritualist.

I saw brightness in Wildman's eyes—he appeared pleased by my inquiry. "Well, it all begins with the cosmos," he began cryptically. He mentioned something about the moon, and then went on to say that very few people understood "what Steiner really intended to say about farming" and how it was, in fact, unrelated to alcohol.

Not related to alcohol? I wanted to know more—I'd always thought of biodynamics in relation to winemaking, specifically.

Wildman shook his head. "Biodynamics," he said, "is a much bigger philosophy about how to live ethically and healthfully, not just a winemaking approach."

I was entranced by his accent, which sounded like I might imagine a character out of a Graham Greene novel would—educated but not necessarily posh, and from an unplaceable time in the past, not our current moment.

"I used to be more involved in the biodynamics community in Australia," he said, now sounding tired. "But it got too capitalistic."

Then he smiled. "And you?" He asked me what my experience was in the world of natural winemaking.

I began telling Wildman about myself, my work as a freelance wine journalist, and my plans to move directly to Paris in a few weeks.

"Well, Paris has the best natural wine bars in Europe, besides Copenhagen," he replied. Just then, the bus slowed down—nearly an hour had passed while we'd chatted. John announced that we'd arrived to the qvevri maker's studio. Out the window, I saw a grassy field with massive clay vessels lying strewn and a large shed of some sort, plus a rectangular structure—a kiln. Behind it all, there was a hillside with a small stream running through.

Zaliko, a sturdy man in his late forties, walked us around his studio, letting John translate his Georgian. He showed us the slow technique by which he constructed these thousand-liter egg-shaped clay vessels over the course of months, entirely by hand, slapping the clay on layer by layer. We poked our heads into the brick kiln where he fired the qvevri at 1,200 degrees Celsius before lining them with beeswax. Zaliko gestured to the verdant hills behind the creek, from which he harvested the soft beige clay for these massive yet elegant containers. The air smelled clean and wet. I blinked back exhaustion, fascinated by what I was seeing.

These qvevri would be sent out around Georgia, and thanks to the recent popularity of Georgian wine, some would even make their way to Northwest Italy and France. They would be buried and used to ferment skin-contact white wines and red wines.

The qvevri was the world's oldest known winemaking technique, and as more producers had discovered how it provided a perfectly cool and protected setting for wine to ferment without sulfites, it was enjoying a fervent revival. Winemakers today have all sorts of technology at their fingertips—reverse osmosis machines, temperature-controlled stainless-steel vats, computer-controlled fermentation, high-speed filtration devices—but there were people who preferred these handmade clay vessels. The old, in other words, had become new.

Chickens were squawking frantically, running around on the dirt outside the shed where Archil Guniava made wine. Archil, a tidy and modest man in his fifties with kind eyes flanked by profound crow's feet, wearing jeans and a collared T-shirt, waved us over. We stepped around the birds (one of their sisters was probably being prepared for our lunch) to enter his shed. It was tidy and dark, with a gravel floor in which only the glass-covered lids of underground qvevri could be seen.

Archil kneeled and pulled a lid to one side. Then, using a large ladle, he reached into the hole to extract wine of a dazzling orange shade. This was poured into each of our glasses to taste.

Archil spoke, and John translated.

"Some of these qvevri belonged to his grandfather," he told us. "Only natural wine has ever been made in this cellar, throughout the years.

In Soviet times, it was illegal to sell anything but bulk wine. People went to prison for breaking those rules. Around seven or eight years ago, Archil began bottling his own wines. He uses the local 'Imeretian method,' with only around 15 percent of the grapes fermented on the skins, but no stems." Despite my experiences tasting Georgian wines back in New York, I was having trouble following the grape names as Archil explained one wine after another, dipping into the underground qvevri for tastes of his most recent vintages. I was familiar with the four most ubiquitous Georgian grape varieties. These were the ones the Soviets had permitted to be made commercially during their seventy-year domination, which involved state control of agriculture, and they included three white varieties, Rkatsiteli, Mtsvane, Tsolikouri, as well as the inky-red, almost black Saperavi. But Archil was making wine from a few of the hundreds of esoteric varieties found only in Georgia: "Krakhuna" and "Tsitska" were two names I jotted down, approximating their spelling. I hoped to absorb more information over the coming week—this was a lot to take in on our first day.

Archil's twenty-year-old daughter, Nino, who had begun helping her father make wine, stood nearby, also tasting. She had bright, dyed-red hair, a thoughtful expression, and like her father, wore jeans. Later we learned that Nino was starting to make her own wine, joining a small but growing legion of independent female winemakers in Georgia.

The sight of Archil withdrawing wine from underground impressed me. "How does he clean the inside of the qvevri?" I murmured to Chris Terrell, the New York importer who'd helped me get an invite on the trip. Chris nodded to a corner of the wall, where a large bouquet of dried leaves on a long stick was hanging from the ceiling. I'd never seen a winery so low tech. Archil told us he used a mechanically operated basket press. There was not a single electrically operated machine in the shed. Yet the system was quite well thought out: the clay qvevri, lodged underground, provided a naturally cool environment, ideal for slow fermentation and aging. They didn't impart flavor onto the wines, as oak was known to. It was winemaking at its purest.

I took a moment to savor a white wine made of a grape called Krakhuna, which I'd never tried before. The juice had been macerated on its skins for four months. Ever since that bottle of Matassa back at Reynard, I'd been falling more and more deeply in love with "skin-contact" wines, made through longer macerations on skins, versus directly pressing freshly picked grapes (as most modern white wines are made). Also called "amber" or "orange" wines, this technique was traditional in Georgia, and it had become increasingly common around the world over the decades, thanks in part to two pioneering winemakers of Italy's northwest region, Friuli: Stanko Radikon and Josko Gravner. Both had, over time, become very well-known practitioners of the skin-contact style, which had a historical form in their region: Pinot Grigio made on its skins, known as "ramato." However, both men had to fight an uphill battle to establish themselves in the late '90s, when skin-contact winemaking was seen as primitive and inelegant across Italy. They both persisted in this vision, and Gravner went one step further. He had learned about Georgian qvevri and slowly began acquiring them over the years, until finally he was able to use only these vessels to ferment his wines, for total emulation.

From across the small winery, lit only by sunlight streaming in through the one door, I watched Wildman and his companion, Tom—the two seemed inseparable. Wildman wore a ratty T-shirt with some Japanese writing on it, along with rumpled jeans and boots, and he had a notebook and pen out. Tom, who was listening intently, wore eyeglasses and had his brown hair, with thin streaks of gray, pulled back in a bun. Like me, they were first-time Georgia visitors, completely absorbed by our humble surroundings.

It was long past noon when we wrapped up the tasting. Nino and Archil beckoned us to follow them to the house, just beside the winery, where Archil's wife, speaking loudly in Georgian, herded our group of fifteen around a large table for our very first Georgian supra. One by one, the dishes came out. My stomach rumbled fiercely, so I bit hungrily into the fried cheese bread known as *khatchipouri*. The bread was soft but had been charred so that it had beautiful black blisters and just a hint of crunch, and the tangy cheese was immensely satisfying. Clearly, the cheese bread and our entire meal had been freshly prepared, from scratch. Soon, our table was piled high with sautéed wild flowers, tender grilled meats, and creamy, homemade cow's milk cheeses. We all clinked our glasses of orange wine and yelled out in Georgian, "Gaumarjos"—Cheers! I snuck another look at this Wildman and found him quite handsome. There was an interesting sparkle in his eyes.

The meal was everything we'd been promised—there was too much delicious food, there were toasts, and inevitably, the cha-cha came out. We'd had plenty of wine at this point—nobody had been spitting, as wine professionals normally do, during the qvevri tasting. And we were all decently jet lagged and underslept. But to say no to cha-cha would have been impolite. Everybody threw back a few shots, and then Archil and his family filled some plastic bottles to the brim and gave them to us as parting gifts—two for each bus.

In the front of our bus, Tom and our host, John, chatted away during the hour-long ride, passing a bottle of cha-cha between them. I saw that Wildman, again seated near me, had finally loosened up. "Would you like a drink?" He offered me the bottle over the seat, along with a gentlemanly smile. I took a swig, then passed it to Irene, the red-lipsticked Italian winemaker, who then shared it with Alex, the bar director from Brooklyn. We all began talking comfortably, the strong grape brandy serving as an icebreaker.

It was nearly dusk when we arrived at the home of Ramaz Nikoladze, a stout, burly man. As many natural winemakers will do with visitors, he insisted that we see the vineyards right away—or else, he said, we could not understand his wines. Soon, we were standing ankle deep in spindly grass and flowers, amongst thick vines that had no trellising to support their sturdy branches—"bush" vines. Fifteen years ago Ramaz planted this vineyard, he told us, and he did something very unusual—he did not cultivate it, at all, instead leaving it to grow completely free, per nature's guidance. This vineyard held a special energy that, looking around me at my fellow visitors, I dared to believe we all felt, simultaneously. A buzz in the air, something unquantifiable. I knew it would be expressed in Ramaz's wine.

"Hey, Anton," called out Tom—using Wildman's proper name. "What's this plant for?" He held up a sprig of something green plucked from the earth.

Wildman looked embarrassed at the attention, but he replied, "That's equisetum—also known as horsetail." Ramaz nodded, confirming. Wildman also said, to anyone within earshot, that biodynamic growers often made a "tea" out of this plant to spray on their vines—along with nettle and yarrow—to prevent mildew. As dusk fell, we made our way back to the house and climbed up a flight of stairs to a wooden terrace, where a long table had been set for dinner.

"I wish I hadn't eaten so much khatchipouri at lunch," I said to Wildman as he and I held out our glasses for Ramaz, who was pouring a white wine from a bottle with beautiful Georgian script on the label.

Wildman nodded and took a big slug. I laughed—he was so irreverent about drinking this wine from such a special vineyard. He shrugged. "Natural wine is the best cure for jet lag. You've got to drink a lot to push through." Then he added, "Plus it will wash down all that cheese bread, whatever you called it."

We were asked to take our seats. I tried not to overthink it as I sidled into a chair across from him. Two people, however, remained standing: John, who was smoking near the balcony's edge, and Tom, who was pacing anxiously with a glass of wine, which he wasn't even drinking.

Ramaz's mother and sisters came out holding trays of food. But I couldn't eat yet—Tom's jitteriness was putting me on edge. I could see others felt the same way. We were all visibly and audibly drunk—you could tell by the cadence and pitch of people's voices—but Tom seemed extra perturbed. He was just standing there, awkwardly, kind of swaying. Behind him, John blew smoke into the forest.

"Is your friend OK?" I asked Wildman, who was helping himself to the plate of roasted potatoes in front of us. I sipped the wine with pleasure but didn't bother deeply analyzing it. I'd turned off my "journalist brain" at this point—it was too dark to take notes, and I was fully overwhelmed by the long day and lack of sleep the night before.

Wildman curved his neck and looked well at Tom, who was mumbling to himself, gazing aimlessly through his eyeglasses. "Tommy? Yeah, he seems to be all right. Hey, Tommy, you OK over there?"

Tom sort of bounced over to the table. "Yeah, I just don't think anybody's having any fun right now. You guys having any fun?" He said the second part loudly. "Is anybody actually having fun here?" Even louder.

At one end of the table, the crowd from Barcelona and France, who seemed to already know each other, was starting to giggle at this drunken performance. But a couple from New York seemed like they were having trouble eating with Tom hovering over them. Ramaz and his family came through with plates of whole grilled fish, and Wildman seemed interested in the food and unconcerned about Tom.

But Tom kept going. "None of you—you're not having any fun. Especially you New Yorkers, hey—you won't even get off your phones!" Now he was full-on yelling, and everyone could hear him. Our hosts looked perplexed, and they stopped serving.

Tom's ranting continued for three or four minutes, with the same themes: nobody was having fun or paying respect, we were all too obsessed with our phones, especially the New Yorkers. Perhaps Tom had been watching us take photos with our phones all day, cringing at every capture. I'd certainly taken my fair share of iPhone shots and had whipped out my digital camera a few times also—I had an article assigned for the website Munchies on *Vice*, so it was justified.

"Can you please do something about your friend?" one half of the New York couple hissed to Wildman. He put down his fork and said meekly, "Hey Tommy, whyn'tya come eat? There's a seat for you down there." Tom glanced toward the empty chair at the end of the table but didn't move.

Instead he yelled, his voice threatening to crack with emotion: "Do any of you even know where we are right now?"

Someone replied, perfectly, "Do *you* know where we are?"

Tom blinked and swayed before announcing with full lung force: "WE ARE IN THE HOUSE OF RAMAZ!" There was a ripple of laughter across the table. Wildman finally jumped up and spoke to Tom, and he seemed to be heading toward his seat, when—

A sound like ripping wood came through the air, and we heard Alice the writer scream, "John!"

Our very tall and well-built host, finishing his cigarette, had leaned too far against the railing and careened all the way down, about three meters.

And so that dinner ended abruptly, with John and Alice in a taxi, making a mad dash to the hospital. We were definitely having fun, I thought as Tom accepted a glass of water from someone, stunned by what had just happened. But I thought I knew what he'd meant—he was overwhelmed by the humility and generosity of these people and felt ashamed that he could offer nothing in return.

Maybe because he was a winemaker, he felt differently. But wasn't generosity a fundamental part of natural wine? Even one glass could be such a gift, as it came about through so many years of hard work.

The next morning, John returned from the hospital, stitched up, and Tom apologized to everyone in the group, one by one. Wildman couldn't stop poking fun at him on the bus. It seemed to put Wildman in a great mood, actually, that his friend had made such an ass of himself. We collectively decided to lay off the cha-cha and focus on the wonderful Georgian wines.

But while I was enthralled by each winemaking family we visited over the next few days, I also found myself focusing on Wildman. On each long bus ride between the remote producers, he told me cheesy (and sometimes dirty) jokes that had me rippling with laughter. I shared stories about visiting natural winemakers in the Loire Valley, and he responded with dispatches from the Jura, where he'd been once or twice. We discussed whatever we tasted at each winery, comparing wines from one place to another and interrogating their differences. He seemed to value my opinion. But frequently, in between the studious talk and hilarious moments, I caught a glimpse of something wistful or angry in his facial expressions.

Midway through the week, on a stretch of road where our bus fell quiet as nearly everyone drifted off into sleep, Wildman and I chatted about our lives back home. I wanted to know more about the origins of the Lucy Margaux Farm.

"Well, the farm started out as a small purchase for planting a vineyard. Then we bought the house, and . . . it's just me now," he told me in a voice heavy with sadness.

Wildman's wife had walked out abruptly, asking for a divorce, only months earlier. It happened right in the middle of vintage. He seemed like he was still in shock.

"How long were you married?" I asked.

"Fifteen years," he replied in a solemn tone.

I had never been anything close to married, and my childhood had been marred by own parents' separation, so I found it difficult to empathize with Wildman's situation. It definitely sounded lonely—his daughter was away at boarding school, so he was living by himself on what sounded like a very large, isolated farm. And I had the impression that Wildman was only beginning to reckon with this significant rupture in his life.

But we quickly shifted out of the serious talk and returned to flirting intensely: Brushing arms. Exchanging generous grins. Inside jokes about who was more nerdy, with our heads buried in our notebooks at every tasting. Sitting in close proximity on the bus—it felt like I had a high school crush and we were on a weeklong field trip. Adult summer camp—or a reality TV dating show, natural wine version. The tension between Wildman and me grew noticeably with each passing day, to the point where our colleagues nudged each other and rolled their eyes at how inseparable we'd become.

But the trip was nearly over, and Wildman and I were still technically platonic. Looking at him drinking a glass of wine at lunch, I admired his thoughtful way of tasting and asking questions, and thought about our moments of closeness over the past few days. I felt that *something* needed to happen, or I'd always wonder—what if?

On our last night in Georgia, we gathered at the restaurant that John owned, located in a small village not far from his Pheasant's Tears winery. By now we were all friends, and the conversation sounded a mild roar in the stone building as we took our seats for the supra. The long room itself had brick walls, with archways carved out revealing a lounge, whose floors were covered in woven rugs. At the start of this special meal, we were treated to a session of enchanting, traditional polyphonic Georgian singing, featuring John's wife, Keto, and a few of their friends.

We passed around bottles of the white, skin-contact, and red Pheasant's Tears wines and helped ourselves to steaming plates of lamb stew and khatchipouri. It felt like a feast from days long past, centuries ago. Now Tom was not complaining about people not having fun or being on their phones. Georgia and its devoted form of hospitality had wiped us all clean of whatever worries or anxious urban habits we'd shown up with. We had arrived in Georgia as ambassadors of our respective natural wine communities around the world, and nearly everyone had bonded over these eight days.

Dinner was over, the party was becoming more raucous, and we were all standing around in a wine cellar. John had opened a few precious older vintages of Georgian wine and wrapped them in paper bags, and people were loudly and helplessly trying to blind taste the wines. Our week of exposure to grapes like Chinuri, Aladasturi, and Kisi was not quite enough to cement their profiles in our psyches, but we guessed anyway, for the fun of it.

Stepping over to Wildman, I lightly tapped him on the shoulder. He swirled a flaming orange-pink liquid in a glass while he chatted with an importer from Barcelona and when he paused to look over at me, I asked, "What are you drinking, can I have some?"

He blinked. "Oh, this is just some fucked-up booze I made." Wildman nonchalantly poured me a glass of his own Pinot Gris, which he'd imported in his luggage.

His casual, almost disinterested stance was the challenge I'd been waiting for during the entire trip.

"Thanks, looks delicious," I said, holding Wildman's gaze intently, to which he responded with confused, flickering eyes. Clearly, this guy needed a more obvious invite. "I might go have a look at the stars, they're really nice right now." I smiled enticingly. "You should join me." It was a cliché, but how else was I to get this unsuspecting man alone? Anyway, he seemed to hardly register my move, and merely smiled blankly.

"Yeah cool, be there in a moment," he mumbled, before resuming the conversation.

I flounced out to a patio built of gray stones. Sitting with my knees tucked into my chest, I watched from afar as the pleasantly drunk crowd smoked cigarettes and laughed at each other's stories, emptying glass after glass. It was no surprise to me that people were drinking excessively. Natural wine, as it's generally low in alcohol, lighter in style, and free from added preservatives, can be frighteningly easy to drink—which means that a gathering like this results in many, many bottles emptied. Everyone wanted to enjoy our last Georgian evening. The stars shimmered above. It was nice to be under a foreign sky, farther away from home than I'd ever been, but I couldn't stop my mind from drifting back west, to Paris. In a few weeks, I would be opening the large windows of some Parisian apartment and stepping out onto a balcony to gaze at the red-tiled rooftops dotted with smokestacks. I took a deep breath, feeling the thrill of not knowing exactly what would come next in my life.

The reality, though, was that I had no such glamorous apartment waiting for me in the French capital. I had a basic-looking Airbnb lined up, on the edge of the city. In my bank account, I had maybe $2,000. I'd invoiced for all of my recent work and finally hustled that one apathetic media company into paying their overdue fees, and I crossed my fingers that the other publications would pay soon. I knew a few people in Paris from my various tours of France and a few people from New York who now lived there. And I was counting on Gaba's arrival in a few months. If Gaba could get over to Paris fast enough, and if we could scrape together some financing from our New York contacts, maybe we'd open our wine bar within a year. I visualized it—we'd find a vacancy near the canal in the gritty 10th arrondissement, where real estate was still manageable. We'd offer a marble bartop where people could order plates of charcuterie and raw-milk cheeses. We'd have a neat selection of French wines, a few Italian bottles. And of course, we'd be pouring my new favorite: Lucy Margaux.

"You're still here!"

I looked up, startled. There was my strange crush, smiling at me, also with surprise. Had I been sitting there for some time? I regarded Wildman's mussed-up hair, already receded to a point where it now encircled his forehead like a crown; his eyes seemed to be a burning shade of blue, they were intensely clear. He wasn't particularly tall or heavily built. So many of the stereotypes I'd acquired over the years of a "handsome" man did not apply here. I knew, as well, that there was about a decade or more between us, which felt like maybe a few years too many. But he was funny and charming and brilliant, and he made incredible natural wines.

The biggest problems I saw were that Wildman was recently divorced and lived in Australia. Which also meant that, no matter what happened that night, surely I'd never see him again, or at least not anytime soon.

"Come have a seat," I said. Wildman obeyed and lowered himself reasonably close to me so I felt the warmth from his body. He cleared his throat.

"How do you like my fucked-up booze?"

I smiled at Wildman's self-deprecation and finished the last of the Pinot Gris in my glass. "It's really stunning, not fucked-up at all."

We spoke for another thirty seconds about the week we'd just experienced, and then I could not handle the small talk any longer.

"Would you like to kiss me?"

He blinked and for a moment did not move, and then he exhaled deeply and leaned back.

"I don't think it's a good idea."

My insides twisted together. How could I respond to that? I nearly stood and walked away. But then I noted the emotion in his voice: this was a man who had been deeply hurt, and recently.

"I . . . understand," I finally said, trying to smile and seem confident, like I was very cool with being rejected. I shifted my posture, again about to stand.

But he put a hand on my arm. "Wait. What am I saying? Come here."

That kiss. It was warm. And not short. It felt like we'd known each other for a long time. It was a kiss that occurs only once, a fire that burns for a specific moment. When we pulled away, we heard Tom calling to Wildman that their taxi had arrived to bring them to Tbilisi. Their flight back to Australia was to leave in a few hours.

We said goodbye with brief words in gentle tones, and then Wildman heaved on his black rucksack and gazed back at me for a moment in silence before turning away. I saw him join Tom, and then they were gone.

I stayed looking at the stars for some time, still feeling his lips on mine. I would have bet you a case of wine that I would never see this Wildman again. If that had been so, I would have always remembered that kiss.

– Four –

Soif

A few weeks later, I arrived in Paris with two suitcases. They were stuffed with some light, summery dresses as well as a couple pairs of jeans, and books, including Quebecois writer Mavis Gallant's *Paris Stories* and my ten-year-old copy of Hemingway's *A Moveable Feast*—and not much else. Literally and figuratively, I had left it all behind: friends, mentors, and most of my belongings. I resolved to free myself of those ties to New York and build the life I'd always dreamed of, which motivated me as I dragged my suitcases from the taxi up three steep flights of stairs to my Airbnb in the 20th arrondissement. It was so far on the outskirts of Right Bank Paris that I could hear the constant roar of the *peripherique*, the freeway that circles the city. I noted that I would be sharing a bathroom with other random guests, and there was a moldy, smelly kitchenette in the basement that I dared not use.

The room itself was clean, and I unpacked some clothes from my suitcase and placed my journal and pen on the desk that sat below the window, which fortunately looked out onto the quiet street below. My room was just a place to sleep, anyway. I was in Paris—a living museum and playground—and basic lodging would do for now.

My morning coffee was a milky *café crème* inside a dusty little *tabac* where locals crowded around the TV with their scratch cards, seeing if they had the winning lottery number. Or, for a change, I'd go across the street and sit outdoors on a blue-and-white wicker chair where my espresso was placed on a faux-marble table. That place was run by an Algerian artist who had left her home country when her friends and relatives cast her out after she had an abortion.

At any rate, the coffee was only two euro at either café. And no one in the neighborhood spoke to me in English, so I found comfort in temporarily losing my American identity.

Over the years, I had met a few people who now lived in Paris. Back at Uva, I'd had a French customer named Louis, who was a wine sales rep. Every Sunday night he would come into the shop and ask me to help him choose a bottle for his weekly poker game. Later, when I began visiting France for press trips, he moved home to Paris and was working at Le Verre Volé, one of the original and most iconic natural wine bars there. He took me out on a few motorcycle rides around the city, bringing me up to the rooftop of L'Institut du monde arabe for a panoramic view over the Seine.

Then there was Sam. This was an oddball—I'd actually met Sam on the dating website OKCupid, back in Brooklyn, where he held onto an apartment, though he mostly lived in Paris. He was a travel writer, penning those "36 hours in wherever" columns the *New York Times* runs. We went out on a few dates in Paris and kept in touch.

Now that I was here, I felt hesitant to contact either man, but it was hard to say exactly why. Just days after our time in Georgia, Wildman had called me out of the blue, to tell me that our kiss on the last night there had been truly special to him and in fact the "highlight" of that trip. The conversation had set me alight. So there I was in Paris, definitely single, and yet I didn't feel flirtatious and free.

To make new friends, I began going solo to wine and food events. Little by little, I connected with American, British, and Australian expats who were in Paris to work in its fabulous restaurants and bakeries, or to write, or just to have their own Parisian adventure. Within a week of arriving, I was not speaking very much French outside my neighborhood, nor was I befriending the notoriously insular Parisians, but I was staying busy. The city's daily life itself thrilled me.

"Ah, *oui*, I'll have the quenelles, and the fish," I told the waitress at the small, simple bistro a ten-minute walk from my apartment. Lunch was two courses in Paris, and I loved enjoying it alfresco in the summer sunshine, with a glass of blond beer—my preference when a natural wine wasn't available. This feast, as it seemed to my American palate, accustomed to midday deli sandwiches and salads, cost at most 18 euro.

It felt so human, the French way of living. "Bonjour, madame," shop owners and bakers greeted me, with utmost politeness. I sometimes bought a baguette and a hunk of firm tomme cheese, picked up a bottle of Gamay from Le Verre Volé, and brought it all to the Parc des Buttes Chaumont for a solo picnic. All around me, couples and families drank rosé and basked in the sun. Someday, I too would be sitting here with my partner and our child, in one of the world's most beautiful cities. Would he be a writer, like me? Someone in the wine industry? Something else—an architect? The tomme and baguette mingled with the juicy red wine like a perfect cacophony of sour, tangy, and fruity flavors. I swallowed each bite hoping that the contentment in my stomach might cover up the loneliness wallowing in my heart.

As the equinox extended the rays of the sun until nearly 9 p.m., casting a hazy glow over the rooftops of the city, an epic *canicule* took hold—the heat wave lasted more than a week and made it impossible to do anything but consume liquids. By then, I had developed an evening routine that was very conducive to this task. It didn't matter if it was Tuesday or Saturday, it went more or less the same . . .

On any given evening I would leave the apartment around 7 p.m. with a small purse holding my wallet, phone, reddish-orange matte lipstick (applied just before leaving), and a pouch of tobacco. The warm air and soft sheen of dusk greeted me as I rolled and lit a cigarette in the street. Phone in hand, I would guide my steps toward whichever wine bar I chose to visit that night. I could take the subway, but my greatest pleasure was wandering through the streets, discovering old buildings or adorable bistros and people-watching along the way.

One evening at *apéro* time, as that hour was referred to, I consented to meet Sam at La Buvette, a hole-in-the-wall natural wine bar known for its popular white bean dish, copied at establishments around the world, and for its owner, a notoriously prickly Parisian woman who liked to glare at foreigners (responsible for perhaps 80 percent of her business) over the rim of her glasses while mumbling details about the wine.

I had mixed feelings as I headed out of the apartment toward La Buvette. It wasn't that Sam was intolerable, by any means—on the contrary, I enjoyed my dates with him. But with two major caveats: one, he just didn't get natural wine at all, and I'd grown tired of explaining it to him. "I found a really great cheap Merlot that's certified organic. Does that mean it's natural?" My response was: probably not, it just means the farming is organic, but they could do whatever they want in the cellar, and anyway, what's more boring than a "cheap Merlot"?

More importantly, though—because, as I had learned in New York, an affinity for natural wine could be an unreliable indicator of whether a lover had potential as a partner—the reality was that, as his fourth decade waned, Sam seemed perfectly happy remaining a bachelor. During our handful of dates we'd had fun, but things were pretty casual. We had not once met each other's friends, discussed going on holidays together, or met up during daylight hours. I didn't see things ever deviating from this pattern.

Then there was the fact that I still didn't feel entirely single, ever since that kiss with Wildman.

As I paced Rue Saint-Maur, looking for the butcher shop that served as my signal that I'd arrived at La Buvette, which was so inconspicuous I often walked right by it, I admitted to myself that I was having drinks with Sam as a test, to see whether he could make me forget about Wildman. That strange Aussie winemaker had managed to command my attention from the other side of the world. I couldn't stop thinking about the hilarious dirty jokes he'd told on the bus or the intellectualism with which he'd tasted wines.

La Buvette was packed with tourists and ex-pats sipping on pink bubbly stuff and ogling display bottles that rested on a ledge along the wall. Sam was already equipped with a glass of red wine—he hated white, I knew, and I'd given up trying to show him that white wines didn't all taste the same. He stood and we kissed on both cheeks, and when he didn't offer or make a move to get me a drink, I made my way over to the bar where the owner, Camille, was preparing a plate of saucisson.

I greeted her in French, and to my relief she replied in the same language, even though I knew she spoke English. She served me an aromatic white wine, and I asked for a plate of the beans.

Once seated I asked Sam, "So, how's your novel going?" The last time I was at his apartment, a short walk from the bar itself, there were Post-its everywhere on the walls, full of notes for his manuscript-in-progress.

He told me it was going well and mentioned a new "36 hours" piece he'd just submitted on Marrakesh. "And how's your magazine? When does the first issue come out?"

"It's coming right along," I told him. All afternoon, I had been editing a feature for *Terre*, the magazine I had dreamed up with two friends from New York. I'd come up with the name, merging my Francophilia with our collective interest in sustainability and the provenance of ingredients and drinks.

"I've been working on a profile of this winemaker in Oregon who puts out a few experimental wines every year, which sometimes have flaws, but he does it to prove that academic winemaking isn't the only way," I explained.

Sam's eyebrows raised, our typical intellectual banter commenced, and I loosened up. By the time we were on to our second glass, I was enjoying myself. But I wasn't sure I'd invite Sam to the next destination, where I had planned to meet a friend of a friend, some musician type from the States who was passing through Paris. After only a few weeks of being an ex-pat in Paris, I'd begun to see how easy it was to end up hanging out with other foreigners who were "passing through" for whatever reason.

"Should we have another glass?" Sam grinned at me with tannin-stained lips. He was so nice and so reliable. I knew our evening would progress as previous ones had, and we'd ride on his scooter back to his place, where I'd spend the night. In the morning, he'd eat exactly one half of an avocado with balsamic vinegar while I smoked a cigarette out the window. Then we'd say goodbye until the next time.

"Sorry, but I actually have dinner plans," I replied. Sam shot back that he, too, had a rendezvous later that night. We finished our glasses in silence, and as I picked up the conversations around me, I concluded that there was not a single French-speaking person in the entire place.

During those five months I lived in Paris I never went to the Eiffel Tower, the Champs-Élysées—or many other well-known tourist attractions. I played the anthropologist, ethnographically studying the tribe of natural wine through diligent participant observation. The people I met came from all walks of life, all over the world.

Although the French countryside can often feel like nothing has changed since the nineteenth century, the city of Paris enjoys a vibrant internationalism. I often noticed, during my various strolls through neighborhoods like the 10th arrondissement, multicultural groups of teenagers hanging out at cafés. Life in Paris is famously very public and visible, taking place in streets and markets, and this brings a sort of accountability. There is a famous saying that a French woman does not dress up and groom herself for her husband, or for any man. She does it out of self-respect and respect for others who see her when she is out in public.

Already within a month of arriving to Paris, I felt my inner *parisienne* coming through. Before heading out to the local market on Saturday mornings, I outfitted myself thoughtfully but not ostentatiously. I always said *bonjour* respectfully and *merci, bonne journée* at the end of a transaction. When I found a table at a café patio, I knew to flag the waiter rather than sit there hoping endlessly that he'd bother to come my way. I learned to pick up my daily baguette either before lunch or in the late afternoon and to nibble on the "toe" end as I walked home to prepare the meal.

While I loved these small intricacies of life, I also found it hard to settle in, not having a job that introduced me to any locals. It wasn't easy to make friends. Louis, who on my shorter visits to his home city had invited me to late-night blind tastings at Le Verre Volé or brought me along to dinner parties, seemed incredibly busy working at the wine bar, and now I rarely saw or heard from him. The only thing worse than being lonely in New York, I thought, was being lonely in Paris, where everyone looked like they were having incredible fun, *pique-niquing* in the parks or drinking cheap beers at the grungy bars of Belleville. I tried to be patient—Parisians, especially those who grow up in the city, are known for being difficult to befriend.

I met more and more foreigners. One night at a wine bar I struck up a conversation with Elisa, from Rome and working at a research institute for the New Wave film director Jacques Tati. There was Nicola, an Australian hospitality industry veteran now working at a popular Parisian bakery and lunch spot, who was well acquainted with Tom and Wildman and other South Australian winemakers. I met Jaimie, a freelance writer from New York who shared my passion for natural wine. Gaba, who'd left her job at June and was now working at an Italian restaurant back in New York, was trying to wrap things up there and figure out what to do with all her furniture and her dog, so she could get over to Paris. On some days, I fully believed that we'd open a wine bar, while on others, I wondered if we could possibly muster up the confidence or the capital.

All the while, Wildman had been pursuing me with text messages, sharing his South Australian life through images and videos. They evoked a world unlike any I'd ever known: Birdsong unfamiliar to my ears. Vineyards bordered by yellow winter flowers. I spent a while stalking him on Instagram and found myself staring at a photo of him making coffee on an old espresso machine, with a bar and open kitchen in the background, finally discerning that this was the restaurant he owned, which he referred to as "the shop." It was much more sophisticated than I had thought. In that photo, making coffee, Wildman wore an expression of sincere happiness that I hadn't really seen in Georgia. When I looked at the date of the post, I understood—it was from before his wife walked out on him. I still didn't know very much about all that, although Nicola had filled me in on some of the backstory. Apparently, the relationship had become dysfunctional over the years, in part due to Wildman's tendency to work obsessively in his winery.

Increasingly, the glimpses I saw of Wildman's life in Australia made me think it wouldn't be the worst place to visit. It could be another adventure for me. But not right now; I still needed to settle into my life in Paris.

After leaving Sam at La Buvette, I arrived at Aux Deux Amis to meet the musician friend-of-a-friend. I stood outside and fiddled with my phone, looking for a message from this unknown acquaintance. On the terrace, it appeared that Paris Fashion Week had taken over the bar:

people were decked out in the latest A.P.C. boots, with cheetah-print jackets, plenty of makeup, and highlighted hair for the ladies, and thick denim coats for the men. Last year I had written an article in *Vogue* naming Aux Deux Amis one of my favorite bistros. I mused that my piece must have caused this influx of fashionistas.

My sort-of-blind-date was at the bar. He seemed about my age or younger, wore a loose, cotton button-up shirt, and had a scruffy two-day beard. We kissed on both cheeks, feigning our localism. He told me that he was a music scholar, visiting for an academic conference, then asked, "So, you've decided to live in Paris. How did that happen? It's quite a big move."

"I've wanted to live here for years, because I'm obsessed with speaking French." I laughed at myself, on my second English-speaking date of the night. "But mostly, I wanted to be here so I could write about natural wine." I wondered if he knew what I was talking about. "Do you like natural wine?"

He brightened. "Yes, totally! I've heard so much about this place, I was really glad when you suggested it." I sighed with relief and began to peruse the wine list.

"You must come here, like, regularly?" Musical scholar asked, craning his neck at the wheel of blue cheese on the counter, the mirror where the night's by-the-glass selections were written in erasable white pen.

"Almost too often," I replied. Everything inside Aux Deux Amis was tinted yellow by the naked tube-shaped bulbs that wind around the entire ceiling, forming zig zags above the small area reserved for actually sitting down and dining. We were being jostled by, on one side, a guy wearing a T-shirt reading "ANTI-WINE SOCIAL CLUB" and, on the other, a pair of freakishly tall, lanky women who must have been in Paris on modeling contracts. They leaned provocatively on the zinc bar in their slim shift dresses, taking all the bartender's attention. I waved in vain for a few minutes until finally, the other bartender, a woman, came over.

As dusk turned to dark night, we began drinking.

"This is incredible," he said, looking down into his glass. "What is it?"

I swirled mechanically. I'd grown so used to swirling my wine, it had become something of a tic. "It's Romorantin, a white grape in the Loire Valley. It's a heritage variety. There's hardly sixty hectares of it left." I went on to tell him about the time I had visited the Courtois family, who made this wine, on that trip to the salons with Phil and Daniel from Jenny & François five months earlier.

The bottle went down very quickly—we blamed the heat.

"Why don't we try a light red," I suggested. "A Gamay, or Cab Franc."

For natural wine drinkers, a "light red" wine is the pinnacle of what our taste buds desire—it's peppery and fresh, can be served chilled, and is low in alcohol, but it has more flavor and complexity than a white wine.

We ordered some small plates—a mushroom salad, bread, fish rillettes—to eat standing up at the bar. By the time they arrived, we were on to the light red wine. By asking in my best French, I had managed to procure a bottle from one of my favorite cult winemakers, Aurelien Lefort, who produces hardly ten barrels of sulfur-free wine each year in Auvergne, in Central France. It was an extraordinary Gamay, and I couldn't stop raving about the "energy" in this wine, which my companion also found impressive. I took a few photos with my camera phone—for Instagram.

Our conversation was getting quite blurry. Music, wine, did we even understand each other, despite speaking the same language? It didn't matter—we were out in Paris, feeling alive.

Just before 8 a.m., I awoke to the sunlight beaming into my room, with fresh beads of sweat forming on my brow. Already, my forehead was pulsing. I was totally naked on top of the sheets. Down on the floor, there was a pile consisting of the previous night's clothing and my purse. I reached for my wallet to find receipts from the night before. I wanted to know how much I'd spent. Upon seeing the amount, a rush of anxiety flooded me.

Just as the hangover came in full force, I remembered that this musical scholar and I had shared a brief make out session there on the terrace of Aux Deux Amis. With all of that wine, I wasn't surprised it happened. I was only glad it didn't progress beyond lip action.

Then I saw a new message from Wildman: a selfie he took in his vineyard. He was wearing a beanie and an enormous grin, waving pruning shears at the camera. The accompanying message said, "It's a beautiful winter day here, my love!!!" He was adorable, farming diligently in the fresh air, while I was drinking myself into a stupor to supposedly start a new life.

More sleep? Water? Coffee? Ibuprofen? Those were my options to get me through the day, and the impending heat wave, already in evidence so early in the morning, was not helping.

Those nights, as much fun as they were, made me wonder how much were they about actual enjoyment and how much were they about mindless consumption, showing off my knowledge about natural wine to look cool, via social media prowess? Was I basking in the glory of all that Paris had to offer—or was I desperately seeking meaning, still in pain from my "breakup" with New York City? Even the "dumper" feels the pain of loss when a relationship ends, we all know. Although I wanted to tell the world that I was having the time of my life, part of me suspected that my excessive drinking was an attempt to escape from the unfortunate truth: that I had no idea what I was doing with my life.

Sure, I was submitting articles to various publications, in print and online. I was starting an indie print magazine. I was improving my French, sort of. But most of all, I was running away from discontent.

I busied myself to ignore all this. In the afternoons, once my hangover had faded, I wrote and edited in cafés. The forthcoming inaugural issue of *Terre* was shaping up nicely—friends and acquaintances across the United States had contributed a personal essay about working at a natural wine bar, an interview with a London-based female chef's fashion company, a story on a coffee expert in Brooklyn, and more. Although the magazine's scope was broader than just natural wine, I was slowly working on a feature about Julien Guillot, an intriguing winemaker in southern Burgundy who captured my heart when I first tried his blend of Gamay, Chardonnay, and Pinot Noir, made from one of France's oldest organically farmed vineyards.

My public writing was focused on natural wine, but personally, I sought to understand Paris on a different level. Ever the diarist, I scribbled regularly in my journal about the endless displays of wonder I found there, such as the signpost at the street where van Gogh lived—as well as the sadder realities: the Syrian refugees living under highway overpasses on the city's northern outskirts, the "closed" sign that remained on the Bataclan nightclub, where the tragic terrorist bombing and shooting killed dozens of Parisians.

Ernest Hemingway moved to Paris in the early twenties with his first wife, Hadley. He began publishing short stories and started writing his first novel, which would become *The Sun Also Rises*, while living in the city during its heyday of jazz, sexual openness, and art. The Hemingways lived a bohemian life in Paris, thriving despite an inconsistent income. When the author did get paid, he and Hadley would go straight to the tracks to sip Champagne and gamble. But because Hemingway so enjoyed the café culture, the food, and particularly the wines of France, he didn't feel deprived. In *A Moveable Feast* he depicts the essence of the bohemian Parisian life in its golden age—a time known as *les années folles*, the "crazy years":

> But then we did not think ever of ourselves as poor. We did not accept it. We thought we were superior people and other people that we looked down on and rightly mistrusted were rich. It had never seemed strange to me to wear sweatshirts for underwear to keep warm. It only seemed

odd to the rich. We ate well and cheaply and drank well and cheaply and slept well and warm together and loved each other.

Nearly one century later, it seemed to me that this elegant yet affordable lifestyle had been replaced by expensive neo-bistros and wine bars that sucked money from my wallet every time I left my dingy third-floor Airbnb apartment.

"Bonjour, Madame Signer?" The French pronounced my last name "seen-yay." It sounded nice. "Yes, we have delivery for you," said the voice on the phone.

"Uhhhhh," I replied, fumbling around for some spring water, the thunder in my frontal lobe reminding me of how many gallons of Gamay I'd consumed last night at La Cave à Michel. I hadn't eaten much—or did I scarf down a kebab on the way home? Regardless, my stomach hurt.

"A delivery?" I wasn't expecting anything.

"Yes, madame. Can you descend?"

Curious, I went to the window and peered out while I took a long drink from a warm bottle of Badoit sparkling water. There was a man standing beside a scooter, holding what looked to be a massive bouquet. I threw on a pair of jean shorts, cringing as I zipped them up over my bloated stomach, and a T-shirt. Holding my forehead, I walked down the stairs, also clutching the railing. The delivery man handed me a bouquet full of peonies and roses a shade of pink that I'd never seen before. It was a *really* big bouquet. I brought it upstairs and placed it on the little desk. Then I walked back down to the street to have a bad coffee and buy a vase from the florist.

As I arranged the flowers, my hangover began to fade. Through the haze, I racked my memory for the last time that a man had given me flowers. Or had a man ever given me flowers? It was something I'd seen in movies but had always relegated to the category of "things that don't happen to women like me." To women who live in ramshackle apartments under the freeway in Brooklyn, or beside the freeway in Paris, because that's all they can afford. Who will do anything to keep up their intellectual passions, anything except give up partying, of course. I believed it was my destiny to remain loveless, gradually assuming the likeness of Gertrude Stein, spending days in a faded, torn armchair critiquing the latest books. She didn't need men, at all—in fact, men needed her and sought her support and approval.

But here was this guy from Australia, crashing down on my destiny of endless one-night stands, as well as that of becoming an island of myself in need of no testosterone-bearing human, with his enormous bouquet of bright pink, aromatic roses and peonies, which were transforming my sparse, white-washed room.

As that scorching heat wave finally subsided, the room filled with the flowers' aroma. Waking up to that smell at its strongest point, I was on the brink of tears, moved by their presence, by Wildman's unexpected presence in my life, even with the distance between us. That bouquet felt like my only true friend in Paris. Every morning, I caressed the bright, fully bloomed petals and beheld their prettiness.

Wildman and I began to speak every other day on the phone, updating each other on the small events of our lives. I consulted a map to find Basket Range, the little town where he lived, at the very bottom of Australia, near the city of Adelaide and not too far from the coast. In the clips he shared, I viewed him helping his friends Tim and Monique, of the winery Manon Farm, dig up cowhorns full of manure that they had buried for months, which they would then dynamize and spray on the vines, a biodynamic fertilizer treatment. Perhaps I was one of the only women in the world who found this romantic! In the background, I heard the exotic Australian birdsong.

It was distracting to be mentally removed from the cafés and cobblestone streets of Paris to the wintry forests and vineyards of Australia, but I found myself addicted to this reportage from Down Under.

Wildman called at all hours, and often if I didn't answer, he called repeatedly until I picked up, then sounded nervous and apologetic. I noted a touch of desperation, at times, in his communications. But I also loved hearing that shakiness in his voice because it laid bare his feelings for me, and if it was sometimes extreme, I chalked it up to the difficult emotions of being recently separated.

When Wildman called to see if I liked the flowers—did I like the flowers?! What a question—instead of talking about the sweltering heat and pangs of loneliness I was feeling, I related the good news that I had found an apartment to sublet, in the nearby area of Belleville, for six weeks.

"Well, I've also got some good news," he said. "I'm coming to Paris . . . and I'm taking you on a date. Or, I mean, well, may I take you out on a date?"

I had to bury my face in my shoulder to muffle the laughter. I pictured Wildman with flushed red cheeks. Given the entirely unconventional nature of our courtship thus far, his query amused me. We had met on the back of a bus in Georgia, drinking crude homemade brandy, telling locker-room jokes—was there a need to be quite so gentlemanly?

"I'd love to go out with you, of course," I told him.

We began discussing his plans, and he explained that after Paris, he'd be visiting Edinburgh, London, Slovenia, Italy, and Spain—and, casually throwing it out there, that I was welcome to join.

"Oh! Well, I, um, thank you." I didn't reply one way or the other—although I considered myself fairly spontaneous and free, I wasn't sure about diving into a trip like that.

I thought about Wildman lodging in one of the expensive hotels of Paris. Natural winemakers don't really make tons of cash, I knew from my years of chronicling their lives as a journalist.

"Do you want to stay with me?" I suggested without forethought. I regretted the words immediately. The pause on his end was dramatic.

"I—I mean, if you don't have any place . . . " I stammered.

I'd been acting so cool up until that point! Now I looked like I would sleep with any winemaker who passed through Paris and rang me up. Or I was giving the impression that I was already totally sure of my feelings for him.

"That's a nice offer," he replied without actually saying yes, and we left it open to interpretation. As we hung up, I had no idea whether he would actually be staying with me or not. But just in case, I shot him a quick email with my new sublet address.

I called Gaba and we exchanged updates. She was getting ready to move over the pond. I'd helped her lock in a server job at a small natural wine–focused bistro in the 10th arrondissement, and she was securing an apartment through a website. I envied her access to these things—she had unique circumstances, thanks to her friendly marriage to a French citizen. Meanwhile, I looked regularly at the website of the French consulate back in New York but found no available visa appointments in the near future, and watched my bank account dwindle every day, as I was unable to look for a job with nothing but tourist status.

I wasn't sure how I was going to make this new life work, unless Gaba and I got the wine bar going quickly. But there was no going back to New York now. I was finished with that life.

My new Belleville apartment was perched atop the ancient cobblestone street of Rue des Cascades—the street name, meaning "waterfalls," referred to aqueducts there that were installed in Roman times—which wound around the corner to an overlook above the Parc de Belleville. From that park, the Eiffel Tower and nearly everything else in Paris was visible. In the evenings, groups of teenagers gathered to blast hip-hop music and smoke joints; on weekends, there was a North African food-and-clothing market. Near the apartment, a tiny, unremarkable bar expanded onto the patio, where I spent afternoons drinking cold beer and eating peanuts while struggling to read novels in French.

In mid-July, another flower delivery came, and again Wildman's presence filled the space. This one came with a handwritten note that read, "Put these in your window, so I know which is yours." I sighed like a starlet in a Hollywood rom-com but then I decided that I, too, could play this game, and I plotted my own romantic ploy for his arrival.

On that day, I struggled to function normally, unable to eat my lunch, too jittery to write as I kept checking my phone for news that his flight had landed on time, and finally decided to wait for him at the patio nearby where I could drink a beer calmly and read. As I left the apartment, I grabbed one of the larger roses from his recent bouquet and began tearing off its petals, drizzling them on the doorstep of the building and along the centuries-old street, which was too narrow for even a small Peugeot to pass through, all the way to the outdoor bar.

Parisians in their twenties were gathered in small groups, consuming vast quantities of beer. I smiled, thinking that only a few years ago, I would have been entirely content to do the same, afternoon to evening. But something had changed, in recent years, I had to admit. It boiled down to one simple thing:

At thirty-three, my biological clock had been ticking for some time now, and thoughts of starting my own family had taken root.

I knew that Wildman was recently out of a marriage. We had spent hardly eight days together. But those flower deliveries, and the constant communication, meant everything to me. Even if this flirtation turned out to be no more than just that, a brief and sweet encounter, it felt like the most promising and true glimpse of romance I'd felt in years. In New York, I'd found one guy after another explaining to me why he "wasn't ready" for a serious relationship—or, as with Evan, outright betraying my trust. Wildman's doting courtship thus far, though it had occasionally been overwhelming, had been a welcome reminder that I was worthy of love. Contrary to what I'd long thought, I didn't need to be further along in my career, or more financially stable, or to have longer hair or thinner legs to feel attractive and appreciated.

I had no idea where it would lead, but my instinct said to give it a good chance. At any rate, we'd have fun drinking together in Paris.

As the sunlight began to fade, I waited. My patience was growing thin. I could no longer consume more beer. My eyes were glued to the rose petal pathway I'd left, but I couldn't quite see all the way to my building entrance. Had Wildman arrived and missed my attempt at a romantic gesture?

Antsy and tired of eavesdropping on the French conversations around me, most of which centered on people's summer vacation plans, I abandoned my table and walked slowly toward the entrance to the apartment, following my rose petal path in reverse. Slowly, Wildman came into view: his scraggly hair, standing up straight from the crown of his head; the heavy black backpack, now resting on the ground; his scuffed-up brown leather shoes, ripped Levi's, and T-shirt. When he saw me, he took a step back, watching me approach.

He looked much the same as he had in Georgia—like a funny, eccentric, philosophical man. He also appeared slightly startled. It seemed we both felt unprepared for this moment, so built up in our minds over the past several weeks.

"Hello," I said, softly. "I guess my rose petals didn't work?" I laughed, unsure of what to do, then he moved toward me, and magnetically I was drawn in, and we kissed.

We climbed the three long flights of stairs and then, the little space that had been mine—where I'd shamelessly eaten creamy pasta directly from the pan in my underwear while staring at my phone, or wrestled the most vicious hangover by sprawling on the couch with a bottle of Badoit and making hourly journeys to the toilet—was now suddenly ours. We stared at each other in embarrassment.

"Something to drink?" I gestured to a countertop where a few half-empty bottles sat at room temperature. I hadn't planned this far ahead.

Wildman shook his head. "I've brought something, but I need to chill it down, first. And I have something to give you now." Perplexed,

I sat on the couch while he put a bottle in the fridge, then rummaged through his bag.

He sat beside me, and I smelled the light fragrance of sweat from the day-long journey through the skies. I hadn't even stocked up on a proper wine to greet the weary traveler, and he'd brought a gift. In my defense, I had, in actuality, met this guy just once and had already seemed overeager by asking him to stay with me.

"Don't you want a shower first?" I blurted out. I was nervous. Wildman simply handed me a plain, cardboard shoebox.

"If it's a flower vase, you're too late, I already bought one," I joked. I placed the box on my lap and lifted the lid carefully. Inside, wrapped in tissue, there were two small glass tumblers that looked to me hand-blown, and a stunningly elegant, sturdy pair of scissors, like nothing I'd seen before. The blades were curved and made of high-quality metal. I felt their weight in the palm of my hand.

"Those are for pruning, in the winter. They're from Japan." Wildman smiled mischievously. "In case you fall in love with a winemaker."

I had understood his romantic gestures when the flower deliveries had come, but a pair of pruning shears was an unanticipated tool of seduction. To no great surprise, Wildman was the kind of man who traveled halfway around the world with a backpack full of T-shirts and a box of bespoke glassware and viticultural tools.

"These are all so beautiful," I said. We hugged.

"Those are made by hand in Adelaide," Wildman said, referring to the glassware. "We serve wine in them at my restaurant."

He said he would in fact take a quick shower. While he did that I stared blankly at his strange gifts and wondered what to make of it all. Maybe I'd gone too far, telling him he could stay at my apartment.

Once out of the shower, Wildman went straight to the fridge and came toward me holding the chilled wine bottle he'd brought with him. I knew from the crown cap and little bubbles that it was a pét-nat.

When he put the bottle on the table, I gasped. There, on the hand-drawn label, was me. It was from a photo he'd taken in Georgia, during one meal toward the end of the trip, when we'd sat beside each other and laughed the whole time.

"Is that—I mean—are you kidding me?" Maybe I sounded unappreciative. But it was a shock to see myself on a wine that Wildman had made, a wine that was presumably shipping around the world, being served in wine bars in Tokyo and Sydney. Wildman acted bemused as he poured out two glasses of the pink bubbly wine.

"To us," he said, clinking his glass against mine, which I held out dumbfounded. The wine was perfectly bright, light, and refreshing, with a blast of acidity—everything I'd always loved in a pét-nat, since those first glasses after my shifts at Reynard.

We downed the wine fast, and Wildman poured another round. Meanwhile, he told me more about the trip he had planned for the next few weeks: first, a natural wine fair in Scotland; then a week of sales in London; eating at Hiša Franko, a famous restaurant in Slovenia; a wine festival in Spain. It sounded exciting, but also frantic. I was weary of airports after a year of back-and-forth travel between Europe, New York, and elsewhere.

All of this intensity was making me fidgety. "Come on," I said. "We're going out. It's Paris! We can't just stay in my apartment!"

We walked as quickly as possible. "What's that bar, where they serve the mayonnaise eggs, and that really sexy couple runs the place, and there are no seats?" Wildman was bad at names, but I knew exactly what place he was talking about, and I steered us up toward La Cave à Michel.

We peered inside and saw that the owners, Ioulia, a petite Russian woman with fierce blue-gray eyes, and Romain, who was French but had Ukrainian heritage, were alone, smoking and wiping the bar. Romain's long, stringy hair hung over his eyes as he kept his head down, tidying. It must have been a slow evening; they were surely getting ready to hop on their Vespa, which was parked outside, to ride home—when they glimpsed us.

"Heyyyyy," Romain came outside, slapping Wildman on the back to bring him in for kisses on both cheeks. Apparently they were friends, from one of Wildman's previous trips to France. Romain disappeared and then emerged from the cellar with a bottle, and proceeded to cook us a few dishes on the induction burner behind the bar, while we chatted with Ioulia. We stood and drank and ate *ouefs mayo*, served in a tall metal structure that held three eggs perfectly, and a dish of mayo served alongside—their unassuming version of the dish was iconic in Paris. We left, arm in arm, and just as I wondered if Wildman was feeling tired, he turned to me.

"Where next, my love?"

It was a spectacular night out, and we'd gotten drunk enough to facilitate the awkward moment of getting into bed together—despite being nearly complete strangers and having had no physical contact other than a couple of kisses. In the morning, we hugged and kissed some more. Then my writer's mind snapped to attention, and I sprang out of bed and began dressing. Wildman lay there, watching me, with a relaxed expression on his face.

"So, I actually have to do a few things this morning," I explained—I needed to finish edits on two major articles for my magazine. Although it was not yet August, if we were going to release the first issue in October, the design process would need to commence in about two weeks. None of us—my cocreators or the designer—had ever made a print publication before, and we didn't know how long it would take.

Wildman's presence in my apartment confused me. He probably felt like coming to Paris was a vacation, but I had plenty to do. How was I supposed to stay on task now that he was here?

He rose and began dressing while I made us a French press in the kitchen. When he entered, I had my laptop open to the files I needed to edit. One piece was looking all right, but the other would require a solid hour of work and I could not put it off.

"Coffee's there," I said, pointing.

He poured his and drank silently while I typed. A few moments later, I felt his hand on my back.

"Why don't I run out and get us something to eat?"

"Great idea," I said, breaking my gaze at the screen long enough to find my keys and a canvas tote bag. "Here, take these. The Bio c' Bon is just out the door, up the stairs that look like they're from medieval times, I think they are, and to the left."

He kissed me and was gone. Forty-five minutes later, he returned, wafting in the scent of a freshly smoked cigarette and holding a full bag of groceries.

"Hi," I said. "I'm sorry about before . . . I just get stressed out, sometimes." I explained that I'd been doing freelance articles, writing about natural wine for various publications, along with preparing the magazine, and it felt like too much sometimes. "Especially with a hangover . . . this is brutal! How's yours?"

Wildman laughed, "Not great. I think a healthy breakfast will set us right."

I kept typing while he began washing and peeling and slicing. In a short time, I saw him going toward the wooden table in the living room bearing plates piled with colorful items. I followed the scent of fresh herbs and saw that he'd prepared a feast: poached eggs with soft cheese and tomatoes cut in jagged shapes so that their juicy flesh nearly sparkled in the late morning light, cucumbers layered with toasted almonds and topped with a dusting of what looked like seaweed, hunks of fresh peaches drizzled with honey. We sat and ate, not speaking for a few minutes, and as the food entered my body I felt instantly better. I thanked him as if it was no big deal, that he'd made exactly the kind of nourishing food I needed, laid out in such an attractive yet simple way, but I felt deeply cared for.

"If you're done with your work, I was thinking, why don't we rent bicycles?" Wildman suggested. It was a warm day with blue skies, and I couldn't think of anything better. Wildman washed the dishes while I put together a bag of cutlery, napkins, and the tumblers he'd gifted me. We walked to the nearest city bike station and rented two cruisers, which we mounted and directed toward the Seine. Wildman rode ahead of me, and I smiled at his ratty T-shirt and longish, uncombed hair.

Somewhere around Bastille, we found a cheese shop, and I bought a few hunks of Saint-Nectaire, a tomme cheese from the Auvergne region; Wildman grabbed a baguette. We threw them into his bag, and I led the way as we crossed one of the bridges to the Left Bank.

In the Jardin du Luxembourg, we cooled off in the shade under a tree and drank a bottle of white Burgundy from a cult natural winemaker named J. J. Morel. It went well with the cheese; they both had a nice balance of richness and acidity that merged sumptuously on our palates.

"So," said Wildman with smiling eyes. "How many dates did you have to cancel to make room to see me? Or did you even cancel them? Should I not assume you're free tonight?"

I played it back to him. "Well, that depends. I'm sure I could be convinced to clear my calendar, if you have something to tempt me with."

We carried on like that, but then Wildman brushed my hair from my forehead, and I saw the look in his eyes. This wasn't a joke. He had come to Paris to see me. What I'd felt on that bus, driving around Georgia, he had felt, too.

"You're very sweet, you know," I said.

He held my gaze, and I felt his fingers graze my hand. "I've thought about you every day since I got back to Australia."

This brought back my urge to laugh, because he'd certainly texted and called me enough that I could have guessed I was on his mind. But we'd joked enough in Georgia. It was time to find out what we could be if we allowed ourselves to be serious.

"Come here," I told him, wrapping my arms around his shoulders.

But then I watched him freeze up as he leaned toward me. His lips pressed against mine, cold as a dead fish. What had happened? Was it because I'd said something nice, from my heart?

We pulled away from each other, and I pretended to be interested in a pigeon who was plucking away at a corner of our baguette. That kiss had not been warm or sexy or fun. And something had to be done. Otherwise, what was the point of all this? He had shown up with a strange gift that almost felt like an offering to go help him clip vines and then hadn't made a move since he arrived.

"OK, stand up."

Wildman blinked, then obeyed, and soon we were standing, facing each other. He was just above my height.

I took one of his hands and squeezed it. "Now," I said, "I want you to kiss me, but really kiss me. Forget yourself. Forget all of this." I gestured to the trees and people around us. I heard Wildman take a deep breath.

He trembled palpably, but that kiss was better. I told him so.

"One more time," I said. "And close your eyes completely, keep them shut."

And there it was—the same kiss we'd shared in Georgia, under the stars, after that final celebratory meal. His heart beat from his chest into mine as his arm wrapped around my waist, pulling me close.

We separated just slightly, and he looked into my eyes, searching.

"Was that OK?"

I nodded and whispered, "That was very good."

Whatever emotional trauma Wildman had suffered in the winding down of his marriage had certainly been significant, and it was still fresh, but I felt optimistic after that kiss, that it could probably be undone. There was a new man in there, and this moment in Paris was for us to set him free.

There was a new woman stirring in me, too. But I wasn't sure I could welcome her fully, yet. Because the man who was calling her to life actually lived in a different hemisphere.

– Five –

On the Road

Outside, the streets of Shoreditch remained cool, calm, with only a smattering of late-night walkers, coming or going from sleek cocktail bars, dingy pubs, and lovers' abodes.

Inside the red-brick housing complex, in the one-bedroom apartment we'd rented for a few nights, we, too, were quiet—yet internally, we were fuming.

"Wake up." Was Wildman *poking* me? Only an hour ago, we'd drifted off to sleep. I'd basically tagged along with Wildman and Tom and their London importer, bopping from one restaurant to the next, as they showed their wines, and it had been an exhausting day. Now this man was rustling me awake.

I reluctantly opened my eyes to his pained grimace. He was propped up on one elbow, looking down at me unhappily.

"What . . . what is it?" I grumbled, blinking into the sliver of moonlight that entered the room.

I'd known as we went to bed that something was off. It had been a strange evening, and when we'd returned to our Airbnb, I'd noted how Wildman's shoulders hunched forward, and he was sullen, not very talkative. We'd lain awkwardly in bed, not touching, and said goodnight as if we had been married for ten years, rather than having leapt into a passionate romance only two weeks earlier.

"Why did you talk for so long with that guy, the owner of the bar?" Wildman whispered now into the dark. I sighed in disbelief. The well-known proprietor and I had been speaking about mundane, professional matters—our friends in common, my magazine, his recent visit to the Rhône Valley; it's what we called *schmoozing* back in New York—when a server approached and interrupted.

"Excuse me," he said to me. "But your friend Wildman would like to buy you a drink."

I flushed with embarrassment. This felt like the equivalent of a man lifting his leg to mark me as his territory.

The owner graciously spoke to the bartender, who was polishing some glasses, and moments later I was being poured a glass of grower Frederic Savart's Champagne, which for some wonderful reason was open. The bubbles perked me up.

Ten minutes later, the owner was going on about the mezcal he was bottling in Mexico and hoped to launch soon in the UK, when the waiter approached again. My "friend" wanted to know if I needed another drink.

I gave the waiter a frustrated look and said I was enjoying my glass of Champagne, thanks, and didn't need anything else. Our conversation resumed, again, and about ten minutes later, Wildman and Tom, who had met us a few days earlier in Edinburgh, came down from their tasting with the sommelier, ready to go to their next event. A taxi was standing by, but I smiled warmly and said that I'd walk—it was a beautiful day, and I was enjoying the atmosphere of Hackney. Also, can a woman get, like twenty minutes to herself, in a day of following around a guy while he works? I took my time as I strolled over to the Laughing Heart, a restaurant focused on natural wines where Tom and Wildman were hosting a ticketed dinner. The night proceeded wonderfully, with course after course of exquisite food paired with my companions' Aussie juice.

So why, now, was Wildman waking me up in the night to listen to his anxieties?

"If an open relationship is what you want," he fumed, "that's not going to work for me." I blinked in exhaustion. Wildman and I had literally been together, if we could even call it that, for all of two weeks. If this wasn't a red flag I didn't know what was. We hardly had any relationship yet, open or closed, and if this type of jealousy continued then we would have neither. For the next hour, we tried to find a resolution, defending our individual actions and feelings with angst in our voices. Nothing felt right. As much as I tried to navigate his feelings, it was difficult for me to empathize, as I knew little about why his marriage had ended—I wasn't sure he fully understood it, in fact—and had only the vaguest of ideas about long-term relationships, at all. I finally turned over, my back to Wildman.

"I need to sleep."

"Fine, I understand. I knew this wouldn't work out."

I ignored this ridiculous comment and drifted off, overwhelmingly dismayed. Things had been going well until that moment. And now it seemed like another potential relationship was probably ending before it even began. I was beginning to suspect that all of this was happening far too soon on the heels of Wildman's separation. And yet, I didn't want to back away. I still felt excitement at being with him. There was more to this than a fling, I hoped.

When I woke the next morning, the bed next to me was empty—except for a piece of paper with a poem written on it. With the sound of the shower running in the background, I read it, ambivalent. Was a hastily written apologetic poem enough to smooth over the wrinkle in our good times?

Wildman came out of the shower with a towel around his waist and sat down beside me.

"I'm sorry," he said. "I was childish. Things haven't been easy for me lately. I have work to do, I know, to improve myself. Will you accept my apology? Can we enjoy the rest of our time in London?"

I was more than willing, although I opted to let him and Tom go off on their own that day. I made my way to the Tate to absorb some modern art, stopping for breakfast at Borough Market. London's Tube stressed me out, so after the museum visit I set off on foot. During the long walk across the Thames, I told myself that I needed to give Wildman the benefit of the doubt. Divorce was a nightmare—I'd watched my parents go through an ugly one. Maybe, in accepting Wildman, emotional injuries and all, giving him some understanding and time, I could help him to heal. And perhaps, by developing my sense of empathy, my ability to be patient, I'd do myself some good in the process.

After all, I had been feeling pretty broken, too, in my own ways.

The trip had started out like a fairytale. Or rather, like a Harry Potter tale.

We'd arrived on an easyJet flight from Paris to Scotland on a gray afternoon. The next day, we walked up a hill and past Edinburgh Castle, which sat solemn and high above the town atop ancient volcanic rock. It was raining—does summer ever come in Scotland, I wondered? Wildman and Tom—who was fresh off the plane from South Australia—were participating in the first iteration of a public wine tasting dubbed the "Wild Wine Fair," held at a local restaurant called Timberyard, which Wildman had said was extremely impressive. Never before had Edinburgh had its very own natural wine fair.

Timberyard was buzzing as the event staff put things in place. Wildman and Tom began setting up their joint table. They would represent the Aussie contingent together, in a room otherwise full of European wines. Meanwhile, I headed for the spot where an importer was pouring Jean-Pierre Robinot's Chenin Blanc pét-nat, "Fetembulles," to get my insides warm. I noted that Wildman was going to be pouring the pét-nat with my own visage on the label. Nobody knew it was me—our little secret.

From as far as London and around Scotland, people had come to taste these wines, pushing their thick-rimmed glasses up onto the bridge of their noses and asking all sorts of questions. Timberyard was a new restaurant whose young chef-owners, Jack Blackwell and Jo Radford, had worked at Noma, which had pioneered a particular brand of locavorism, sourcing virtually every ingredient from the nearby land and sea, often through diligent foraging, using fermentation and pickling to extend offerings beyond the warm seasons. They, along with India Parry-Williams, their partner in Wild Wine Fair, had not only imported this approach to their kitchen at Timberyard; they had caught the natural wine bug during their internships in Copenhagen and imported this ideology to Edinburgh. Apparently, the city was quite thirsty—Wild Wine Fair had sold out of tickets; the place was full to the brim within the hour.

I tasted widely, hardly bothering to spit: Claus Preisinger's smoky Blaufrankisch from Austria's Burgenland and Yann Durieux's expensive sulfur-free Premier Cru Burgundy wines went down easily. I tasted with two of my natural winemaking heroes: Antonio di Gruttola and Daniela De Gruttola, the couple who run Cantina Giardino, a winery in Campania, Italy, that made my favorite skin-contact wines, which I'd discovered on the shelves at Uva years before.

Sometime in the late afternoon, when the room seemed to be swaying with the tipsy energy of the attendees, Wildman and Tom moved some bottles aside and jumped atop their table. Tom pulled a piece of paper out of his pocket and began reading—it was a poem, cowritten on the merits of natural wine. He passed it to Wildman. Everyone listened, entirely amused by these Australians with their six-day stubble, unwashed and ragged hair, and ratty T-shirts.

"Natural wine is not about capitalism," read Wildman in one verse. "It's about creating more beauty in the world."

When the poem ended, everyone hooted and applauded, and we went back to drinking. It lasted well on into the night, and when we finished all the bottles open at Timberyard, we moved on to the city's charming bars, drinking beer when our palates could withstand no more wine.

For a woman with Italian-Jewish heritage, the austere, rainy beauty of Edinburgh was stretching the limits of what I might call a romantic getaway.

"Is this *normal* for August?" I huddled close to Wildman to stay warm as we strolled along a touristy street, the afternoon after the fair.

"Not exactly summery," he agreed. He wanted to show me a "fancy" shop with "real Scottish clothes." Wildman, who had visited Edinburgh before, seemed comfortable here, whereas the place felt very foreign to me.

"What is your background, exactly? You were born in Johannesburg . . . " I pried, wanting to know more.

"My mother was from Germany," he said. "My father was of English heritage." Having left South Africa as a family to emigrate to Australia in 1984, during the height of apartheid, Wildman never returned.

The shop had men's kilts, women's tartan suits, and plenty of tweed and plaid vests that Wildman was threatening to buy. I felt I needed to distract him immediately, or else I'd be hanging out with a man wearing a plaid vest over a wrinkled T-shirt for the remainder of our trip.

My eyes fell upon an army green, cotton blazer. "This would look really nice on you," I said. "Do you even have a jacket with you here in Europe?" I urged him to try on the blazer. As he shrugged his shoulders into the coat, this one item quickly transformed my Wildman into a dapper gentleman. Still rustic, but with a polish.

He wore it out of the shop, onto the cobblestone streets of Edinburgh. We stopped into a pub for a late morning pint. Sitting on the patio, tourists passed by our table, wearing cameras on their necks and dirty Asics sneakers on their feet.

I still had questions for my new boyfriend. Wildman had mentioned that he might look for a gift for Lucy, his teenaged daughter. "So, how old is Lucy?" I asked.

He leaned back and tilted his head upward, searching for the correct answer. "She's . . . sixteen?"

The uncertainty was comical. "And she lives with her mother?"

Lucy was presently at boarding school, he explained, but hated it and would be leaving when the term was finished—and yes, she would live with her mother, in Adelaide.

We chatted more about our families: I explained that I had three older siblings, who each had two small children. And maybe the large Guinness was going straight to my bloodstream: I asked the question that you generally do *not* ask a man whom you've been seeing for hardly two weeks, who is fresh out of a relationship.

"Would you ever have another kid?"

But he didn't balk. Instead, he sat straight up, then relaxed back into his seat—attentive, yet cool. "Honey," said Wildman with complete confidence, looking me in the eye. "We can have kids anytime you want. So don't even worry about that, OK?"

Too shocked to reply, I laughed awkwardly. But something in me that had been on guard began to ease. Thankfully, Wildman changed the subject—to wine.

"Should we drink this Poulsard?" He gestured to his leather hip-bag, where we'd earlier stashed a bottle of Julien Labet's stunning light red wine from the Jura region of France, along with a wine key and some glasses we'd grabbed from our Airbnb.

A few moments later, we found ourselves sitting on grass in a cemetery where, reportedly, J. K. Rowling had strolled while writing Harry Potter, finding inspiration for her subjects on the gravestones. With the sun creeping out in the late afternoon, and the violet-hued, rose-scented Poulsard in our tumblers, it didn't feel strange at all to be having a romantic moment amongst the deceased of Greyfriars Kirk.

The next day, we boarded a plane to London, where Wildman and Tom would be "working the market." Within minutes of taking our seats, I pulled out the novel I was reading, and Wildman plopped his head heavily onto my shoulder. It sat there, a weight that distracted me increasingly over the two-hour flight. I was tempted to shrug him off me, but he looked so tired. Pouring wine at the tasting had meant having maybe two hundred conversations in one day. His snoring proceeded at a volume that I thought even the pilot could hear. I finally gave up and closed my book for the duration of the flight. When we landed, Wildman woke up and stretched luxuriously. The flight had clearly been great for *him*.

In London, I then saw that my sleep was going to be disrupted for reasons besides snoring. Our late-night argument in that Shoreditch flat was enough to tell me that being with Wildman was not always going to be smooth sailing. But I wanted to stick it out. It's not every day that you meet a biodynamic winemaker on the back of a bus in Georgia who sends you multiple bouquets of flowers to your Paris apartment, then flies across the world to take you on a whirlwind trip. And it's not every day that he indicates he might want a future with you, the kind of future you've been hoping to build but that had eluded you until then.

If you'd asked me where Slovenia was prior to August 2017, I would have squinted and envisioned a country bordering Austria and the Czech Republic—in other words, I would have been thinking, wrongly, of Slovakia. It's a sad truth of being an American that in high school, we only vaguely learn about the part of the world where Middle Europe meets with the former Yugoslav nations. Perhaps that's because the history of that broad region is incredibly complicated.

It's also a region that has been slow to open up pathways with western Europe. Case in point: I had to fly to Venice, Italy, to get to Slovenia from Paris. After London, I had returned to Paris just for a weekend, to move out of my beloved Belleville sublet. I'd deposited my suitcases at the apartment of Nicola, my Australian friend who knew Wildman and Tom well. She had offered to host me upon my return from traveling with Wildman, for the remainder of August and part of September, until I went to work the harvest in the Loire Valley.

Once I'd moved in, Nicola and I headed over to the brownish waters of the Right Bank canal. I brought along a bottle of Dominique Derain's Chardonnay that I'd picked up during a research trip in Burgundy earlier that summer, as well as some stinky Époisses cheese, and of course, a baguette. All along the banks, there were hordes of Parisians enjoying midsummer.

Nicola told me how her job was going at a bakery, and I filled her in on the trip to Edinburgh and London with Tom and Wildman. I really needed her perspective on things.

"Honestly," I began to complain, "there were times I thought I was going to go crazy. He was so clingy, and his emotions were all over the place. I wanted to get to know him and have fun, but it ended up being really intense."

Nicola wrinkled her brow in thought. "I know what you mean. His wife left abruptly, although I think we all saw it coming, in hindsight.

Actually"—her tone shifted to encouraging—"I can really see you and Wildman together. Give it a chance."

The advice was what I'd wanted to hear. And so, instead of resuming my life in Paris, I got on another plane. Before meeting up with the boys, I treated myself to twenty-four hours in Venice, where I consumed a plate of spaghetti drenched in squid ink, caught a Philip Guston retrospective at the Accademia, and snuck into an orchestra concert inside a stunning centuries-old church. Then I waited just outside the lagoon for my wild Australians, who were arriving from Tuscany. I was ready to wilt in the late morning heat when I finally saw the Fiat 500 approaching, with the top down. There was Wildman and his goofy smile, shirt off—Tom's was unbuttoned all the way. Even though Wildman and I had been apart for only three nights, I had missed his sense of humor, his warmth, his somewhat unplaceable accent. He yelled out, "Ciao, bella," and a sense of well-being spread through my body. I couldn't wait to have his arms around me.

We crossed over the border to western Slovenia, which was breathtaking—verdant hills, the gushing Soča river, and not a single commercial development in sight.

Our first stop was, perhaps unsurprisingly, a restaurant: Hiša Franko. Although it was award winning, I'd never heard of it. The owners had come to Australia the previous year for a food event and dined at Wildman's restaurant, and now he wanted to check out their turf. We pulled up to what looked like a large cottage, with a courtyard as well as a few tables and dozens of green plants under an outdoor covering, and found a receptionist. But he seemed puzzled.

"There's no reservation here, for three people, under that name," he told us.

I looked at Wildman. Since Venice, he'd been allegedly sending text messages to Walter, one of the restaurant owners, updating him that our alleged two-person reservation had become three. He asked now as to the owner's whereabouts, and we learned that he and his partner, head chef Ana Ros, were away for the weekend. Walter hadn't seen Wildman's texts. We had no reservation.

Fortunately, the guys talked us into a table for the night, which wasn't easy, given that Hiša Franko had recently been ranked on the influential "World's 50 Best Restaurants" list.

There was also, unsurprisingly, no place booked for us to sleep. Tom and Wildman stood for a while outside our rental car, pondering what to do as the afternoon grew late. Taking charge, I pinpointed the nearest town, Kobarid, and directed Tom to drive us there. We arrived to spectacular views of the snow-capped Alps. As we wandered between hotels, trying to find a vacancy, I noticed various signposts referencing Ernest Hemingway, who apparently had written much of *A Farewell to Arms* in Kobarid, surrounded by the front lines where thousands of men had died in the Battle of Caporetto, a decisive moment in World War I when the Austro-Hungarian forces pushed back the Italians.

I found a wrinkled, barely acceptable dress to wear—I hadn't had time to do laundry during my weekend in Paris. Wildman and Tom put on, to my surprise, collared shirts. We made our way to dinner after what had turned out to be a long day, and upon sitting down at one of the outdoor tables, we immediately asked for some natural wine. The young, clean-shaven sommelier went away and returned with a funny-shaped, clear bottle. It was narrow on top, bulbous and large on the bottom. "This is a Chardonnay from a very respected producer here, called Batič," the sommelier said, pouring for us, adding, "biodynamically farmed."

We tasted eagerly, smiled broadly, and then as soon as he walked away, began to criticize the wine in unison.

It was biodynamic, which was fantastic for the earth, and a great start in terms of making a natural wine. But, as Wildman put it, "too much sulfur." Tom and I agreed. The wine tasted tight and confined, and not living, thanks to the addition of sulfur. Wildman guessed that it had "fifty parts," meaning fifty parts per million. We politely pushed our glasses to one side and asked the bewildered sommelier, who had expected us to love the wine, for "something really crazy."

We wound up with a bottle of Franco Terpin's Sauvignon Blanc, shaded bright orange from extended skin contact during fermentation. It had strong aromas of fresh peaches and clementines, and more important, it had edginess—it was so alive, it even seemed like it was on the brink of being faulty. It was made without any sulfur.

Franco Terpin wasn't technically a Slovenian winemaker—he was based just over the border in Friuli, in a village called Collio, which historically was deeply connected to Slovenia, and only part of Italy since 1945, when Slovenia joined Yugoslavia.

"He's near Saša, I think," said Tom to Wildman, who nodded.

"Saša . . . as in, Radikon?" I asked. Tom nodded and explained that they had met Saša Radikon, a famous natural winemaker, at a festival called Rootstock in Sydney a year or two earlier. His wines had been an early inspiration for the two of them. "Sam showed them to us," explained Tom. I didn't know yet who Sam was, although the mention of his name seemed to trigger sadness in Wildman and Tom. But I definitely knew about Radikon. When Tom suggested we try to visit him during our time in Slovenia, I wholeheartedly agreed.

In the world of natural wine, there are a few names that inspire absolute reverence; Radikon is one of them. In New York, these highly allocated, mostly skin-contact white wines are difficult to come across and expensive to enjoy. A Radikon bottle was striking in that it came in two unique sizes, either 500 mL or 1 L as opposed to 750 mL. But despite the great branding, Radikon wines weren't just status symbols. They were completely unique in taste. A glass of Radikon was powerful, electric energy.

We finished our lengthy meal with a fragrant dessert—crumbly walnut cake with cow's milk kefir and pollen ice cream, alongside pear poached in chamomile and drizzled with local honey—then drove through the dark to the little hotel in Kobarid and slept.

The world was spinning, inside my head. My gut was spinning, in the back seat of the Fiat 500. The guys chatted happily in the front, while Tom drove like a true Italian, taking the curves at breakneck speed. One moment, we were in Slovenia, the next, a road sign told us we'd passed into Italy. Then around the next bend, back to Slovenia—and so on. I no longer saw the natural beauty around me—I only concentrated on holding in my breakfast.

We were late to visit Saša Radikon because Wildman had decided to stop and chat with the owner of Hiša Franko earlier that day, and then we'd all gone for a quick skinny-dip in the river. Now we were rocketing through the hills—incidentally, *collio* is Italian for "hilly"—at the expense of my personal equilibrium.

We arrived in Oslavje, the valley where the Radikon family's home, vineyard, and winery has stood for several generations, and Tom and Wildman lifted themselves out of the Fiat 500. We'd driven with the top down, and Tom had wisely pulled his long locks into a bun, but Wildman's hair was sticking straight up. I opened the door and spilled out onto the grass, where I lay belly-up for the next twenty minutes, taking deep breaths.

Once I recovered, I followed the voices until I found Tom and Wildman standing with the larger-than-life Saša, just outside the winery. Saša, this icon of natural winemaking, paused to look at me from his height of six feet. "I'll be right back, I'll just ask my mother to bring you some of our grappa," he said in a booming voice. "Then you'll feel better."

We headed underground to the winery, where large, wooden fermenters filled the cramped space in which the famed Radikon wines are made. Saša explained the winemaking approach: three to six months of maceration on the skins for the white wines, then around two years of aging before bottling. Finally, Saša's mother, Suzana Radikon, appeared with a small glass of strong grappa, which I swallowed. The dark cellar came into focus, and my headache began to subside.

It is sometimes said about natural wine that "the wine makes itself," or that "nonintervention" means doing nothing. Nobody thinks wine makes itself. What is meant by the term "natural wine" is that nobody *forces* fermentation to happen by adding yeasts, no sugar is added to boost alcohol, no flavor or color adulterants are used, no filtration is applied, and no fancy "reverse osmosis" machines are deployed to engineer the wine. All of those practices are quite common, in commercial and even boutique wineries. At Radikon, by contrast, only organically farmed grapes and, sometimes, small doses of sulfur went into the wine, and the techniques were hardly modern.

This approach originated with Saša's father, Stanislaus, known as "Stanko"—a man whom I'd heard spoken about in tones of reverence amongst the previous generation of natural wine professionals in New York. In the 1990s, Stanko was part of a group of pioneering winemakers from the Italy-Slovenia nexus who decided not to pursue the conventional winemaking that was all the rage and instead committed themselves to old-fashioned techniques, organic farming, and minimal or no sulfur additions. Radikon was (and still is) often spoken about alongside another now-famous producer, Josko Gravner, known for his exclusive use of Georgian qvevri in the cellar.

Whereas modern white winemaking eschews skin contact and generally relies upon the newly introduced stainless-steel tank for fermentation—think of a sharply acidic Italian Pinot Grigio for reference—Radikon and Gravner embraced these antiquated forms of winemaking. In an "a-ha moment" that for Stanko began with the area's native grape, called Ribolla Gialla, it was found that leaving white grapes on their skins for months at a time before pressing delivered an entirely different wine. As Saša recounted for us, Stanko had loved the strong aromas, bright orange hue, and rich texture of the skin-macerated Ribolla Gialla, and soon all the Radikon estate white wines were made this way.

The large open-top oak fermenters in the cellar around us were used during this skin-contact phase, Saša explained. "The grapes get destemmed, and we punch down the cap a few times each day during fermentation," he added. It was a simple wine cellar, but it produced some of the most exciting wines I knew, and not everybody was invited to view it with Saša himself.

"Your wines, and your father's wines, they showed us that sulfur-free winemaking was possible," said Tom, laying his admiration bare.

In 2016, when Stanko passed away from cancer, his loss was felt in New York and other natural wine capitals. By then, Saša had already been working alongside Stanko for several years and had even debuted a new line of wines, one featuring Pinot Grigio and a blend called "Slatnik," both of which were bottled earlier than most Radikon wines and received a small addition of sulfites. Saša hoped these bottlings would draw in new fans and make the overall Radikon range more approachable and accessible.

After touring the winery, we headed into the newly built tasting room upstairs, to taste from the recent release of bottles, alongside slices of pecorino cheese. The wines tasted alive, as vibrant as the Merlot just entering *veraison* just below us on the hill.

The next day, we made our way back to France, flying into Toulouse and renting a van. After stocking up on several hunks of cheese and a few bottles, we awaited the arrival of a train from Paris, which carried Chantal, an American friend of mine. Although Chantal's line of work was running a photography agency, she loved natural wine and was staying for several months in Paris, like me, to explore it more. We'd bonded a few weeks earlier over dinner and a bottle of Partida Creus wine at the seafood restaurant Clamato. When I learned that her boyfriend had recently abandoned plans to join her in Paris and that she was feeling—like me—lonely and confused, I invited her to join us.

Chantal hopped into our van holding a small tote bag of clothes and nothing else, and we headed off to the Spanish border. We'd rented a house for two days on the Catalonian coast.

Wildman and Tom, in the front seat, burly and unshaven, were frothing with excitement. "Yeeeeeow!" sang Tom as he pulled the van out of the city onto the freeway—I laughed, remembering his awful performance on cha-cha in Georgia a few months earlier. It occurred to me that this trip to Europe was Tom and Wildman's primary way of cutting loose, amid a year of constant hard work in the vines and cellar. Tom had a young daughter and wife at home, both of whom he mentioned regularly, but here, he was unhinged.

"Let's crack open a bottle!" Wildman made the suggestion, handing back a wine key.

"Definitely the Leroy," I responded, reaching in between my legs for our stash. We had been lucky to come across a wine by Richard Leroy at the bottle shop in Toulouse. Leroy was a winemaker in the Loire Valley whose bottles were nearly impossible to find, they were so limited in production. He made two bottlings of Chenin Blanc and nothing else.

Chantal and I swigged the precious elixir directly from the bottle before passing it up to the front, where Wildman and Tom were sharing a plastic cup. Tom, as our driver, agreed to drink modestly, though none of us particularly cared. Wildman put on Fleetwood Mac's *Tusk*. When it ended, we stopped for a lunch of rice and seafood at a café in the town of Perpignan. When we finally crossed over into Spain, many hours of driving later, it was dusk.

Seeing the Mediterranean Sea, we parked the car and stared out at the crashing waves. Then Tom said, "Well, what are we waiting for?"

We stripped off all our clothes and ran straight into the frothy water, screaming at the top of our lungs. Teenagers on the beach cheered at our naked bums. If Chantal had been new to the group only that morning, by now, several bottles and a nudist swim later, she was part of our strange little road family.

Our purpose of visiting Spain was not just to frolic in its heavenly waters. We were there to attend H20 Vegetal, an annual natural wine fair organized by two of Catalonia's leading producers, Laureano Serres and Joan Ramon Escoda. Both Laureano and Joan Ramon have long been known as two of the loudest and most fervent partiers in the world of natural wine. They called their festival H20 because they believed natural wine should be like water: refreshing and easy to consume, so that it could be drunk in large quantities. For Laureano and Joan Ramon, the natural wine life meant collectively giving the middle finger to fancy, expensive Bordeaux meant for years of aging in a well-stocked cellar.

On a steaming hot afternoon, we drove through terraced groves of olive trees, heading inland, until we arrived at Pinell de Brai, the small town where Laureano's winery was located and the site of the tasting—or festival, really.

"Ugh, we're so late," I complained as our van rolled into the small town, following our GPS. It was hot, and Chantal and I fanned ourselves dramatically, while the men were shirtless. We had spent the morning slowly cleaning up the rental house, then Tom helped out a friend who needed a ride, so now there was only one hour left of the tasting for that day. To make up for lost time, we quickly raced through, greeting and tasting wines with vignerons from France such as Thierry Puzelat, as well as local Catalonian winemakers Partida Creus and Toni Carbó. We ran into people from our trip through Georgia, and everyone roared with encouraging laughter once they figured out that Wildman and I were together.

You can (and generally should) spit when you taste wine in a professional setting—but at a natural wine event, most people are happy to actually drink the wines, especially in the presence of the producers. I was definitely *not* spitting as I sipped Oriol Artigas's Pansa Blanca from north of Barcelona; or the organic, kimoto-style Terada Honke sake from Japan; or the various blends from the Central France–based winemaker Anders Frederik Steen, whose labels make political statements such as "Don't Throw Plastic in the Ocean Please."

The next day, our quartet returned to H20 with our notebooks out, mildly hungover, and tasted properly, taking time to get to know producers who had traveled from around Europe and beyond to show their wares. A few people waved at me—I had apparently met them at the previous night's after-party, which we'd attended at someone's home and of which my recollection was very faint.

After the event closed, Chantal, Tom, Wildman, and I stopped into a deli to purchase tinned fish and melons. We took our picnic and drove the van to some cliffs overlooking the sea, where we opened up a few bottles of wine. As darkness set in, because we hadn't had enough of being naked in the water, we all stripped off our clothes and bobbed around in the salty waves under the brightest yellow moon I'd ever seen.

Some natural wine bars are so iconic, people will move from one country to another just to work there—to be exposed to that many wines, meet their makers, and be part of a well-established scene. In Barcelona, there is Bar Brutal, located amongst the winding streets of the bohemian El Born sector. Brutal is where our crew spent our last night together before the four of us each went home in different directions.

From the outside, it looks unremarkable, but when you enter Brutal, it is floor-to-ceiling decked out with the wildest natural wines from Spain and beyond. On the walls, visitors have scrawled drunken graffiti with permanent markers: *Solo quiero el vino natural!* "I only want natural wine!"—the battle cry of these fanatics.

Our evening began in a civilized way, sitting down to dinner with a few of the winemakers and importers who had been at H20, enjoying dishes of warm olives and cheese platters. By 2 a.m., everyone from the bar was standing in the streets with our glasses full of wine, smoking and laughing. Eventually, we stumbled back to the apartment we'd rented for the night.

Wildman and Tom left the next afternoon. Chantal headed back to Paris; I moved on to visit a few winemakers in Italy for an article I had been assigned. Twenty-four hours later, I stood in a winery in the Franciacorta region just outside Milan, drinking biodynamically grown sparkling wine. Wildman and Tom would just be reaching their destination, Adelaide. My phone buzzed with a text message: "Got back to my farm in Basket Range. Snowing here. Missing you already, my love." I tried to focus on the winemaker I was interviewing, but I felt a dull sense of loneliness coming over me. I wondered how soon Wildman and I would be together again—and where?

– Six –

Vin Nature

About halfway through September, on a sunny morning inflected with just a touch of encroaching autumnal chill, I piled some sweaters, T-shirts, and jeans into an old wheeled suitcase borrowed from my mother. The Metro took me over the Seine, where I exited in the Gare Montparnasse, to board a train for a three-hour ride heading west from Paris to Angers. I thought back on the last time I was there, having a late dinner at Chez Remy with Phil and Daniel from Jenny & François, after a long day of tasting. That night, and my romance with Evan, felt very long ago.

I exhaled as the train left the station, full of anticipation.

"Have a wonderful time, I will be thinking of you, xoxo Wildman," popped up on my phone. I knew by now that if it was morning for me, it was late afternoon for Wildman in Australia. I also knew it was early spring there. We had been talking nearly every day in the recent weeks since he returned to Australia. But instead of replying immediately, I gazed out the window as the scenery turned suburban and industrial.

A moment to myself, first—this was going to be *my* experience and nobody else's.

It had been my dream for a while to work harvest in la Loire. Although many oenophiles would consider Burgundy or Bordeaux the pinnacle of French winemaking, the Loire Valley region was virtually synonymous with natural wine. Listing all the natural winemakers in the region would have been an almost impossible task, because every year it seemed that someone new started their own label and became the latest hot vigneron. At the same time, each year stories reached New York of the vineyards in Touraine and Anjou suffering from spring frost and summertime mildew, thanks to the Atlantic influence, bringing in cool, wet air, often at the wrong time.

I was about to have an immersion experience, in this mecca, during the 2017 harvest. Previously, I had picked grapes one day here, one day there, in California and Burgundy. I knew it was backbreaking work, yet I signed up for two weeks of harvesting. After all, how could I really claim to write seriously about natural wine, unless I knew it from within?

"Salut!" I waved to Joseph Mosse as he crossed the parking lot toward me. We kissed on both cheeks, and he took the handle of my hefty suitcase and began wheeling it over the cobblestones, past a *boulangerie* fragrant with lunchtime baguettes.

"Did you have a chance to eat something?" It impressed me how well Joseph spoke English. I wanted to speak French with him, but I felt trepidation, despite all the classes I'd taken in Brooklyn and having lived in Paris for several months. I hoped that would change after two weeks with his family.

"I did," I said. "A *jambon-buerre* that I grabbed back in Paris." The combination of salty ham and creamy butter had assuaged my natural wine headache.

He nodded. "OK, then let's go to my apartment. My car is there."

Along with his younger brother Sylvestre, the twenty-eight-year-old Joseph (affectionately, Jo) had taken the reins at Domaine Mosse in recent years. Their father, René, who had retired because of lower back distress, was known as one of the founding figures of France's natural wine movement. Réne and his sons and I had crossed paths several times at tastings in New York and wine bars in Paris, and I knew that neither of the brothers quite fit the image of a winemaker. On this day, Jo was wearing his typical uniform of clean new sneakers with jeans and a simple T-shirt. He could have been a lawyer in training except for that mischievous look in his eyes—and for the magnums of Chenin Blanc standing ready in his fridge, one of which we proceeded to drink in his apartment with his flatmates before embarking on the forty-five-minute drive to the Mosse family home and winery. On the way, we chatted about people we knew in the natural wine world, and he told me he was no longer dating a woman I'd seen him with in Paris, an extremely attractive sommelier at the famed restaurant Le Chateaubriand. Jo didn't seem worried about the breakup—his looks and winemaker status would surely bring along new flames in no time.

On my lap, my backpack was heavy with sturdy, waterproof boots and my computer. The first issue of *Terre* was nearing its publication deadline, and we'd be finalizing the design during these two weeks of harvest. I felt the stress of the project in my tense shoulders. But I was determined to be present as much as possible. The magazine was not going to improve by me obsessing over it during this harvest internship. Most of all, I hoped that Wildman would not bombard me with emotional texts, as he had several times in the past few weeks. He was clearly struggling to reconcile the intensity of our time together in Europe that summer with the immense distance between us now. I also was, but I didn't want to miss a single moment of these two weeks in the Loire.

As the car left the highway and rolled through the quiet hamlet of Saint-Lambert-du-Lattay, we passed house after house constructed of simple, gray concrete, with large windows affixed upon the rooftop, heading toward the Mosse family winery. The vineyards were tucked away in valleys behind the houses. I could understand why Jo and Sylvestre preferred to have apartments in the dynamic city of Angers. Living out here would be lonely for these highly extroverted, well-traveled men in their twenties. I thought of wineries I'd visited in Oregon, like Division Wine Co., Bow & Arrow, and Teutonic, who'd chosen to set up shop in the city of Portland, an hour's drive from the vineyards—because they preferred city living to the valley. It made sense.

Jo stopped to buy cigarettes and I accompanied him into the tabac. He insisted on the *bio*, organic, rolling papers. As I would soon find out, just about everything in the Mosse family home was organic—the butter, the veggies, and, of course, the wine.

"So, how are the grapes looking this year?" I asked Jo.

He lit a freshly rolled cigarette and stared calmly at the road. "Actually, we are going to see right now. I need to check on one of the older parcels. And we have already picked a small amount of Gamay and Grolleau." Those two red varieties, the latter being a light red grape indigenous to that area historically used in blending, were happily fermenting in the winery. Now, we would begin picking the signature grape for which the region is known: Chenin Blanc.

Jo pulled the car onto a gravel road, and soon we were flanked on both sides by large, gnarled vines, matured and thickened with decades past. I could see green-hued grapes peeking out from beneath the canopies. As we got out of the car, Jo grabbed something from the glove compartment. I trailed alongside him as he began picking berries from the clusters of grapes.

"You can take some from the other end of the row," he said. I followed his instruction and came back to him with a handful of grapes. I snuck a taste—they were semi-firm and acidic.

Jo smushed up some of my grapes with some of his in the palm of his hand, then put the mash in between two panes on the device he'd brought. He then held up the device, a refractometer, designed to measure the potential alcohol of the juice, and looked into it from the other end.

He lowered it and nodded. "Not quite ready. Maybe another week." And with that, he started walking back toward the car.

Among us wine nerds, we all have certain varieties that excite us—some sommeliers act like a teenager at a Taylor Swift concert when Cabernet Franc is being poured, while others obsess over Pinot Noir to the point you'd accuse them of being in a cult. Maybe there are personality traits that lead us to certain grapes. I have always loved the rustic, unpredictable nature of Gamay, and I also adore the white grape Chenin Blanc—which is a chameleon of sorts. It's said that Chenin transmits the soil it's grown on more than other grapes. In other words, it shows terroir. Chenin can be sometimes floral, sometimes smoky. For many decades, it was made in a sweeter style, but that has fallen out of fashion, and now it tends to be cuttingly dry, expressing more earthy tones.

Jo's car pulled into a driveway just beside a stone wall, which was adorned with the well-known Mosse "logo"—a swirl of gray and red-orange lines in the corner of the family name, as if someone had placed a wine glass down on paper several times and stained it. Ivy had crept over the logo since it was plastered there, probably in the early aughts, shortly after the older generation of Mosses began making wine there. We had arrived at Jo's family's winery and home.

The first time I'd drunk a really spectacular Domaine Mosse Chenin Blanc wine was at the Lower Manhattan wine bar Pearl & Ash, on a spring day when I had the afternoon off from Uva. Two friends from my biweekly tasting group joined me just as happy hour commenced—which meant that bottles were half price. Sitting on the patio, pretending not to hear the trucks roaring by us along the Bowery, we eagerly paged through the wine list, which was organized first by country, then by region.

"Ooooooh, let's get this!" I pointed to a listing for Domaine Mosse, "Arena," 2015—their wine made from Savennières, a very small Loire Valley AOC (meaning, controlled appellation) where only Chenin Blanc grows.

Although my friends worked in the wine industry, too, I hardly let them have a say in the matter—having tasted a few Mosse wines before, I knew this was the bottle for that moment; my instincts spoke clearly. It arrived, and our glasses were poured. The wine was golden, and it shimmered in the early evening light.

"It's that schist," pointed out my friend Jeff. He explained that the schist content in the soils of Savennières was known to trap the heat, making for powerful wines.

That wine had me hooked on Chenin: it was powerfully mineral, almost like wet stones, but also saturated with flavors of wild, semi-tart peaches, like you'd be lucky to find at a roadside fruit stand in the country. It was a wine full of character—lean yet proud, seeming to transmit a sacred union between soil and grape.

I was ready to help make some of the best natural wine in France, to the fullest extent of my (entirely untrained) ability. And I had almost no idea what to expect.

The next day, at 7 a.m., I stood in the cold on the driveway between the Mosse family home and their winery, the steam from my double espresso rising into the morning fog. I'd slept in a small bedroom adjacent to the kitchen in the family house, sharing the room with a young French woman named Nina, a family friend. The other pickers began to arrive, most of them in pairs, on foot, having parked their cars or bicycles on the road outside the winery. They had on beanies or hoods pulled over their heads, and some were smoking. We all mumbled *bonjour* to each other in a tone that said it's too early.

Jo and Sylvestre were in their standard uniform of baggy jeans and hoodies, but they also wore sturdy boots, and the expressions on their faces were serious; after all, this harvest was their family's livelihood and legacy. They busied themselves hosing down the large, modern pneumatic press, which was stationed outside, just in front of the winery. A moment later, I climbed into an industrial gray van. At the wheel was Jo and Sylvestre's mother, Agnès Mosse. She drove, deep in thought. Gray had taken a firm hold in the long braid hanging down her back. I had never met Agnès, and she struck me as a great beauty.

Half an hour later, I was holding a pair of *secateurs*—clippers specially designed for cutting grapes off a vine—and a bucket, gazing out at a sea of vines.

"Allons-y!" yelled Agnès—let's go! She shepherded us into pairs, so that each picker was facing another as we moved along the rows, and handed us each a bucket. My picking partner, a young man with dark hair and sullen, tired eyes, grunted a "good morning" to me and coughed before crouching down and getting right to work.

And we went, grape cluster by grape cluster, vine by vine, cutting the ripe, yellow-skinned Chenin Blanc from the branches they'd been growing on since June. Within ten minutes, my lower back had a dull ache. Within twenty, my knees were crackling. My hands quickly became cold and wet from the dewy grapes. I was starting to notice a lot of murky gray stuff on the grape clusters—it was rot, and I suspected that it was not the so-called noble kind that makes Riesling taste interesting. This seemed like the unwelcome kind of rot that would lend a putrid taste to the entire vintage.

I held up one of the more offensive clusters and showed it to Agnès. Her face held mostly constant, but I detected in her eyes a look of dread.

"Il faut sentir," she said—meaning, smell it, and see if there's any vinegar. She rigorously sniffed the cluster, then dropped it into my bucket. "It's fine." But if there was a smell of vinegar, she said, then the cluster had to be dropped.

Agnès hollered out the new instructions to the entire team of pickers. As I went on, I dropped probably one or two clusters from each vine. The smelling and dropping of fruit were taking up time. Around me, everybody was doing the same. It was tedious, tiresome work.

An hour later, Agnès yelled out that it was time for *la pause*—this was a break for a quick smoke and maybe peeing somewhere nearby. Then we worked for another hour and a half, until finally we could break for *casse croûte*—the morning snack. Jo and Sylvestre drove up into the vineyard in a truck and pulled out a thermos of coffee, a homemade sponge cake, and some stinky goat's cheese. By then, the sun had come out and we pickers lay in the sun smoking hand-rolled cigarettes.

Meanwhile, the brothers inspected the grapes we'd picked. I could see them conferring with their mother about the extent of the rot. Clearly, this had happened quickly, to their surprise. The brothers had been diligently visiting the vineyards, each day leading up to harvest, walking the rows and tasting grapes, to discern just the right picking time. But the family rented and managed fourteen parcels, spread out around the area. Like many newer wineries, Domaine Mosse wasn't one large estate, with the vineyards romantically set on a hill just beyond the house and ancient cellar. Rather, they had accumulated the contracts for these parcels over the past seventeen years, and each site was unique. Even with Jo and Sylvestre routinely checking the alcohol levels, it seemed that the family had missed the ideal picking window for this one vineyard, and rot had crept in. There was nothing to do now except sort out the unsightly clusters and hope that other vineyards would be in better shape when we got to them.

I sat on the ground with my coffee and a rolled cigarette beside a young woman with intensely dyed blonde hair with bangs, who was stroking her slim black dog and smoking.

"Est-ce que vous avez du feu, s'il vous plait?"

She smiled at my garbled textbook French—the better way to ask would have been, more informally, "T'as du feu?"—Got fire?—and handed me a lighter. I learned her name was Marie and that she was from the region and hoped to become a perfumist. She thought working harvest was an interesting way to improve her sense of smell. And she was living in her parents' caravan at the nearby campground. "With her, him, and them," she said, pointing to other pickers.

"Out of curiosity," I asked, "how much do you get paid to pick?"

Ten euro per hour was the answer. For such backbreaking work! I found it hard to imagine that this was desirable compensation. But everyone seemed to enjoy the social nature of picking. There was a lot of chatter amongst the pickers, who all shared their life stories from between the vines—variations on a theme of choosing seasonal agricultural work in order to exist outside the confines of a nine-to-five job. Some of them seemed to be "between things" or just figuring out their paths in life.

By day three, people had become friendly and relaxed around each other. They asked about my interest in wine, finding my profession of wine journalist to be quite exotic, and politely corrected my French pronunciation. One guy, Faroud, was usually singing along to music he played on his phone, smoking a joint in the vines. He'd pass it over to me when I was facing him from the other row, a welcome respite from the lower back pain I'd grown used to.

After our daily casse croûte, it was back on, picking until lunch back at the winery, in a large room dedicated to such meals. On most days, René spent all morning cooking. He usually prepared a big, hearty stew, some version of *boeuf bourguignon* made with meat from an acclaimed local butcher, and a large green salad, for which I was grateful—salad is not so common in France. We'd sit at the communal tables and pass around unmarked bottles of the previous year's wine. Often, the family would slice up some nice hard cheese for after the meal. Then, a quick smoke and back out into the vineyards.

By the end of the first week, my hands were blackened from grape skins—an attempt to wear gloves had been futile, as they'd been soaked by grape juice within minutes. I'd come to see the pain in my back as a new friend that was with me for the long haul, and I'd developed a deep appreciation for the work that families do to make natural wine.

While we at Domaine Mosse—and other small-scale and natural wineries around the world—snipped each cluster of grapes and inspected them for quality, large-scale, often corporate wineries circumvent such efforts by employing machines instead of humans to pick the grapes. Those vineyards are generally farmed using pesticides and herbicides for guaranteed and maximum yield. Once the grapes are hauled in, winery workers douse the fruit lavishly with sulfur to eradicate any bacteria. The grapes continue to receive sulfur additions as well as flavor adulterants throughout the vinification process, resulting in a beverage that is highly profitable and commercially viable but that resembles Coca-Cola more than wine, and that carries nothing in the way of the soul of a place or of a people, as this wine most certainly would.

Although many French vignerons welcomed the arrival of modern technology and oenological chemicals after World War II, as they offered more stable wine and therefore better income prospects, we were going back to the older ways of doing things. Natural and artisanal winemakers wanted to really allow terroir to shine through in their fermentations and deliver quality in the bottle. So, they doted on their vines, using gentler products like copper and sulfur to protect them from mildew, and worked manually as much as possible throughout harvest and processing. Sure, we didn't have a horse to carry the grapes back to the winery, and the press was electrically powered, chez Mosse. But the spirit was definitely about careful, thoughtful production of wine through handcrafting and without chemicals.

On Friday, we picked for seven hours out in the vineyards and returned to the winery at dusk. I took a hot shower, spending a long time with the running water pummeling my sore lower back, then went into the kitchen for a glass of water. Through the entryway into the living room, I saw a hand, lying on the carpet. Slightly concerned, I tiptoed closer to the door. There was Agnès, fully clothed in her tough jeans and waterproof boots, lying outstretched on a yoga mat in what looked like Savasana, her eyes shut, braid draped over her shoulder. Upon closer inspection I saw she was smoking a cigarette, the smoke floating up past framed school photos of her sons as smiling blond preteens, through the simple white curtains, and out the window to the winery, where those same young men were processing our day's pick.

The next morning, we started picking under the cover of hazy fog. Shivering in my lightweight jacket, I approached the vines, which stood like ethereal Louise Bourgeois creations, sturdy and sensual. I felt a remarkable emotional wellness as I clipped the fruits of these distinguished plants, which were sixty-year-old, gnarly grandmother vines whose trunks were as big as my thigh. Old vines are beloved by winemakers for producing elegant grape clusters, and I admired these elders for their strength and vibrancy.

Picking is a simple task, resulting from a simple truth: something is planted, it grows, it bears fruit. The work sucked up my physical and mental energy, and yet my constant muscular pain was masked by the unique beauty of each site. It was a surprisingly mystical experience for me. I relished the feeling of being lost in a forest of vines, with only the birds chirping as our soundtrack—occasionally punctuated by someone, probably slightly stoned Faroud, blasting hip-hop over headphones at a volume that any nearby picker could hear.

Over the weekend, we took a break from picking. The other workers relaxed at their campground, while back at the Mosse home and winery, I volunteered to help Jo and Sylvestre load the press with some Gamay that had been fermenting. Sylvestre's girlfriend of many years, Adeline, was standing ready, wearing jeans and a wine-splattered T-shirt, with her long dark hair pulled back. She had an enormous grin as she watched the boys set up. Jo, driving a forklift, moved a large fiberglass tank in front of the press, which sat on the driveway outside the winery. His brother then took off all his clothes except his boxers and, shivering, climbed up and into the tank. Jo stood beside the tank and handed Sylvestre a bucket. Disappearing from our view, he lowered into the tank, which was nearly as tall as his two meters, and then he surfaced with that bucket full of fermented red grapes. He then handed the bucket to me—I sat atop the press, high up off the ground, with its rotating door open.

I dumped in the bucket, handed it back to Jo, and we repeated the process. All the while, Adeline stood guard, using buckets to catch the free-run juice that flowed rapidly from a valve at the bottom of the tank. This free-run juice, as winemakers call the liquid that emerges due to the weight of the grapes, rather than deliberate pressing, was then poured directly into the press tray below.

After about forty-five minutes, we were done loading. Sylvestre raised his fists in the air and yelled out, "Woop woop!" He looked like he'd committed a murder—purple and red splotches covered his chest and arms. I closed the heavy door to the press and climbed down from the machine, and Jo concentrated deeply as he pushed a series of buttons on its exterior. It began churning, the full chamber moving in circles, while inside a bag applied pressure to the grapes, and soon, juice began to trickle out into the tray below with the free-run juice. And that was it. That was how wine was made: through long hours of picking; just the right time of fermentation on skins for color, tannin, and flavor; some form of pressing; and then aging in whatever vessel the brothers decided to use on this particular wine. A series of carefully measured decisions but, essentially, only grapes and human labor.

As the second week got underway, I felt desperate for a break from picking. Although it was a beautiful experience for a few hours, ultimately it did get both physically and mentally numbing. I asked Jo if I could stay back and help in the cellar after lunch, and he shrugged and said, "Pourquoi pas?"

In much of France, wine cellars are centuries old, dark, moldy, and located underground or dug into a hillside. Those were the romantic cellars I'd been in during my inaugural wine trip to Burgundy a few years earlier. However, the Mosse family essentially made wine in a garage, which they referred to as a "cellar." Inside this garage, I stood around awkwardly for a moment, getting my bearings. There were various tanks made of fiberglass and stainless steel and a section consisting of barrels of assorted shapes and sizes. It resembled a walk-in closet that had been developed over years of collecting tidbits from thrift shops, neighbors, or secondhand online—there was no state-of-the-art machinery in here.

The brothers had Mos Def playing over a portable speaker. In the driveway, Jo was washing some barrels. As he hoisted a barrel off the prongs of a barrel washer, steam rose, and a few liters of lees rushed out like a frothy Slurpee. He put that barrel to one side. Shirtless, Jo's pectoral muscles flexed as he hoisted another recently emptied barrel on top of the washer.

Slowly, I inched closer to the clean barrel. "Should I move this?"

Jo spun the barrel on the washer and grunted, "Sure." In his face, I saw pure exhaustion. He and Sylvestre had been going to sleep around midnight each night, after the last press had finished. In the mornings when I lined up with the other pickers before piling into vans to go to the vineyards, the brothers were already in the winery, doing punch-downs or tasting the previous day's juice.

I put both hands on the barrel—a 225-liter barrel, the smallest size—and pushed hard. It barely rose off the ground. The coarse wood grated my palms. With my knees, I tried inching the barrel forward.

"Owwww!" The barrel had moved, but it immediately knocked right back into my kneecaps. Jo glanced over at me. I must have looked pathetic, grimacing in pain.

Taking a deep breath, I tried to topple the barrel one way, then the other, making semicircles. This seemed partially successful: there was at least some movement, vaguely in the right direction.

"Ahem." Jo was beside me, with his tired face. I stepped aside like a true amateur and watched him put both hands on the top rim of the barrel and then overlap them, one over the other, spinning the barrel on its bottom rim and moving it in seconds.

I thanked Jo for the lesson and managed to emulate it with one or two more barrels, then said I needed to use the bathroom and exited the winery, where I felt like I was being more of a hindrance than a help. The next morning, I reluctantly returned to the backbreaking work of picking.

René, Agnès, and their sons must have noticed my sense of defeat. At dinner that night, while we all blinked to stay awake, René asked if I might like to join him the next morning. He was going to do some shopping in the local market, and then stop by the Savennières vineyard—from which my favorite Mosse wine, "Arena," was made.

I accepted immediately. Then I returned my attention to the wine in my glass. René had poured it from within a black canvas bag, so I had not seen any details of the bottle. This was our nightly ritual—Nina, my roommate, and I joined the family for a simple dinner of pasta or soup and salad with bread, and we blind tasted. The other pickers cooked their own dinners at the campground. They were probably having a great time, but I enjoyed the coziness of the house, and I definitely appreciated the access to René's well-stocked wine cellar. Sometimes, at dinner, the family brought out their stash of nice cheeses—this evening we had a very fragrant Camembert on the table.

It was a white wine, with notes of wet rocks and lemon zest. Judging from the golden hue, it seemed to have some age. Definitely French, since probably 90 percent of the wine in that cellar was French. I ventured a guess of the region: "From the Loire?"

René nodded sharply. He was generally a jokester, but blind tasting brought out his intellectual side.

Jo's face, too, was stern with concentration. But he didn't speak yet.

Sylvestre seemed disinterested in this game, and I saw out of the corner of my eye that he was texting with his girlfriend. Agnès tasted and considered, but her head was rested on one hand with the elbow propped up on the table. It had been a long day for her, managing the pickers, making sure not a single row was missed, that sorting was done correctly and efficiently.

But I wanted to impress my hosts. "Chenin Blanc!" I went right for it.

Jo nodded. "*Moi aussi*, I also think it's Chenin." He made a face of deep concentration. "But not ours."

René smiled, obviously feeling very clever, and topped up our glasses, peering at us from over the thick, black frame of his bifocals. "So then, where is it from?" His question implied we had correctly guessed the variety. I inhaled and tasted again. This was a structured, *accomplished* wine—it had strong, enticing notes of acidity and stone fruits in harmony, and a sleek cleanliness that reminded me of fresh country air. All that had made me think of Chenin, along with the shimmering golden color. Now I had to decide if it was from this region, around us. It was an easy assumption to make, since this was in fact the homeland of Chenin Blanc. However, it would be a nice move to select a Chenin from, say, the Languedoc, to trick us.

Blind tasting is partly about questioning the idea of terroir. If wine can really reflect the place it comes from, shouldn't we be able to guess just by tasting something where it was grown and made? In my experience, it's often very hard to guess the provenance of a wine, but in some cases, it's like the wine speaks directly to you. A Pinot from Burgundy can announce itself quite obviously. A Cabernet from Napa also can—although some might say that's more due to winemaking style than terroir, since Napa is a relatively young wine region.

Either way, Jo and I were able to intuit that this Chenin was indeed from nearby in Anjou. But even once we'd confirmed that, neither of us had any luck getting the producer. As to the vintage, we surmised that the wine was approximately four or five years old.

René stood up and fetched the bottle, and revealed it to us: 2010 "Les Rouliers" from Richard Leroy. Only six weeks earlier, I'd been drinking a younger wine by Richard Leroy in the van with Chantal,

Tom, and Wildman, driving through Southern France toward Catalonia to attend H20 Vegetal. I wondered what Wildman was up to at that moment—probably waking up early to fix his tractor or hand-hoe his vineyard. We had been texting only occasionally since I'd been with the Mosse family.

We all thanked René for opening this bottle, as it was something of a rare bird.

I cut myself a slice of Camembert as dessert—it was René's favorite. Written on the package was "Camembert du Normandie au lait cru, moule à la louche," meaning that it was made with raw milk and ladled by hand, not machine—the only *vrai*, or true, Camembert, René loved to say. Everyone was falling asleep at the table, so we finished our wine, cleaned up our dishes, turned off the lights, and went to bed.

September's winds rustled the vine canopy as René and I walked through the rows, examining the grapes, which were, sadly, few and far between. René had brought along a refractometer to measure the potential alcohol of the vineyard.

Savennières wouldn't be well known without the presence of one particular winemaker, who effectively put it on the map: Nicolas Joly. And it's fair to say that the biodynamic wine movement also owes much of its breadth to Monsieur Joly. His family estate, Coulée de Serrant, is so large and historically important that it has earned its own official appellation. I can't think of another estate in France like this. Cistercian monks originally planted Coulée de Serrant in the twelfth century—a small monastery still stands just before the hillsides of the property's three vineyards. The Joly family took ownership, and Nicolas began to take the reins in 1977 after leaving a career in finance in London. At first, he used herbicide in the vineyards, as was common practice at the time in the Loire. But he felt disturbed by this approach and saw that the vines were not thriving.

Joly began to implement biodynamic viticulture on the Coulée de Serrant property—he was one of the first in France to do this. He began the so-called 500/501 treatments where cow manure is buried for months, then made into a spray for the vines, and he used plant-based treatments instead of chemical herbicides and pesticides. As he saw vine health improve with these natural approaches, Nicolas became a fervent believer in biodynamics, and his influence spread, partly through books he authored, and partly through friendships with individuals such as Madame Lalou Bize-Leroy of Domaine de Romanée-Conti and Domaine Leroy in Burgundy, both biodynamically farmed. Nicolas Joly also founded the Return to Terroir association, which promotes this kind of winegrowing by supporting members who work biodynamically.

Despite Joly's influence on France's biodynamic movement, his aristocratic background does stand at odds with the scrappy roots of many natural winemakers. On the way to Savennières, in the car, René asked me whether I'd visited the appellation before, and I mentioned that I'd stopped by to taste with Nicolas Joly the year earlier. "Oh lalaaaaa," René had said with exaggeration. "Monsieur Joly himself, how special!" I laughed, but I could see his point—the two men had come from very different places, socially. Prior to starting Domaine Mosse with Agnès in 1999, René had owned a wine bottle shop in Tours. He did not and probably never would own his vineyards, as it would have been too expensive to outright purchase them all. Instead, the family had ongoing contracts with their owners.

We arrived to an underwhelming sight. The Mosse's Savennières vineyard seemed, unfortunately, to be producing very low yields—on some vines, only three or four clusters, instead of ten, were ripening. Frost had hit the vineyard two years in a row. René measured a few berries in the refractometer, holding it to his eye, and said that they were close to 12 percent alcohol. He'd wait another week or two to harvest here.

That afternoon, I picked with the crew. They all asked where I'd been, and I was forced to admit, somewhat sheepishly, that I'd been excused from vineyard work, in the name of journalism.

"But you are paid also?" One young woman asked me. I shook my head and she nodded, understanding now that this was just an "experience" for me, to learn from, rather than seasonal work.

We were in a Cabernet Franc vineyard that day, and the grapes were nearly perfect, unlike the Chenin Blanc with its rot problems. The vines here had some weeds around their trunks, and there were even beans planted as cover crops in between every other row—a sign that the owner of this plot was a committed organic farmer. It occurred to me then that I was uncertain as to whether Domaine Mosse had an actual organic certification.

The next time I saw Jo in the winery, I asked about this. He shook his head.

A total of fourteen hectares, owned by several individuals, comprised the Mosse family holdings. I couldn't help but also wonder: "If the vineyards aren't certified, how do you know that all the growers you work with are truly organic?"

He seemed slightly taken aback at the question, and I worried I had offended him.

But he replied in a gentle tone, probably remembering that I had no way of seeing his point of view. "You can just tell," he said, "by the way the soil looks, and the vitality of the vines. And you also trust. We trust our growers. We've worked with them for many years. It's a relationship."

Some people insist that "natural wine," rather than being an ambiguous category, should become a regimented classification, with standards that would include certified organic or biodynamic farming. But for families like Domaine Mosse, the amount of paperwork (and fees) involved in certifying all these disparately owned properties is prohibitive or at least an annoyance they'd prefer not to undertake.

Many natural winemakers don't mind not having organic certificates and are happy to work outside of the appellation. Their wines are marketed as *vin de France* rather than boasting the appellation they are from. For those who would love to display the name of their town on their labels, having the vin de France stamp is lamentable. But for many natural producers, being outside the appellation is par for the course, as they are proudly swimming against the commercial wine tide. Often, appellations, which determine quality through tasting panels, encourage winemakers to add yeasts and other adulterants and to filter, to achieve a standardized profile. Natural winemakers, obviously, don't want to do that. They want to show terroir naked, on its own, through organic farming and as little intervention as possible in the process. To them, *that* is the truest way to let a wine reveal the place where it's made.

One evening, toward the end of my second week with the Mosse family, as I went to shower, Agnès was in the kitchen, and she asked me to join her to go see the new Jean-Luc Godard biopic. The family friend who shared my room, Nina, was also coming. We arranged to leave at 7 p.m., and when we met in the kitchen, I laughed openly—all three of us had put on bright red lipstick and combed our hair, eager to be dressed for the city rather than covered in leftover grape must and dirt for a change.

We drove to Angers to see the film *Redoubtable*, a portrayal of Godard as a troubled French genius. Afterward, we shared a cheese plate and a few glasses of red wine at the bar Le Circle Rouge, itself named for a French crime movie. Just across the street was the restaurant Chez Remy, where Evan and I had kissed and sparked our evening of romance back in February.

"So, how did you like the film?" Agnès asked us.

I said, in French, that I thought the film itself was fine, but I appreciated how it shed light on Godard himself. "I've always admired him," I explained. I counted his films *Breathless* and *Band of Outsiders*, which sparked the French New Wave by merging subtle social criticism with artistic cinematics, among my favorites. Nina agreed.

Agnès replied that, in her day, Godard was seen as a visionary. He symbolized change for the French youth. I couldn't help but picture

Agnès heading to the cinema as a teenager, smoking a cigarette, maybe wearing the same felt beret she had on now.

Back at the house that night, my brain was too awake after the film and conversation, so I stayed up reviewing the design layout of *Terre*. Finally, it would be off to print very soon.

The next day, I was back in the vineyards, but I could manage it because I knew I'd be heading back to Paris in a short time. For the Mosse family, on the other hand, it was now peak harvest, and exhaustion seemed to be taking over. Even René's cooking, normally outstanding, was reduced to a few variations on simple, one-pot pasta dishes. Adrenaline had been coursing through us, but now that most of the grapes were happily fermenting, that energy was subsiding. We were hurting from lack of sleep.

With warm hugs and kisses to all the Mosse family members, I boarded a return train to Paris with a six-pack of Mosse wines as well as a profound headache from fatigue and the demands of bilingualism—and probably also from overconsumption of carbohydrates, meat, and alcohol. As we pulled into Gare du Nord, I swore to never again make a fool of myself by pretending I knew how to maneuver a barrel or pointlessly break my back picking grapes on cold, wet mornings. Winemaking demanded patience and physical strength that I just didn't have. I thought of Jo and Sylvestre's late nights waiting for the press to finish before they could empty and clean it. I pictured Wildman pruning his Pinot vines in the rainy Australian winter. I couldn't really see myself in either scenario. On the patio of a café, working intently on my laptop or reading a novel and dreaming of writing one myself, is where I most belonged.

Nicola needed her couch back for a visiting guest, so I went to stay at a friend's apartment in Montmartre, one of the most historic and architecturally stunning parts of Paris. Every morning, I jogged along the steep, cobblestone hills, dodging tourists around Sacré-Cœur. Now that the first issue of *Terre* had been sent to the printer, my days were spent trying to figure out what to do next. I'd been only intermittently in touch with Wildman during harvest, and we resumed our digital romance, which was fraught with anxiety. He called frequently, often when I was out walking around Paris, and if I didn't answer he would call again and send text messages expressing frustration.

Frankly, it was a little annoying, and as much as I missed him, I was also feeling a little smothered. He seemed to be pressuring me to commit to something I wasn't yet sure I wanted.

My friend who'd loaned me the Montmartre apartment was returning soon, and I had a decision to make: between moving into my own place—difficult to do, given that I was still on a tourist visa—or continuing to vagabond around Paris.

Then, finally, Gaba arrived, and she made it easy for me.

"My apartment is furnished and you can totally sleep on the sofa for as long as you want," she told me when we met up for drinks at Septime La Cave, a popular natural wine hangout. I could hardly believe she was there. I'd nearly given up hope that she would manage to leave New York—I knew it was no easy task. But Gaba was definitely in Paris. She had moved into a shoebox studio in the 10th arrondissement, where two of the city's main train stations made for noisy daily life. Soon, I had my suitcases stashed under that sofa, and Gaba and I followed a morning routine as if we were adolescent roommates in college.

At 9 a.m., Gaba's alarm would go off, waking us both—the apartment was tiny, with one thin wall separating the bedroom from the other spaces. Gaba would put on the coffee pot and shower; meanwhile, I stretched. When she came out in her bathrobe, we had coffee and smoked cigarettes out the window, which was never closed, looking onto a courtyard in the middle of the building. Then Gaba dressed and did her makeup in front of a tiny mirror above the kitchen sink, before leaving for her lunch shift at a bistro. I spent the morning writing freelance articles for publications back in the States and sending out sales emails for the magazine, which would be self-distributed via personal connections my partners and I had built up over the years. I also finally managed to get an appointment at the French consulate in New York for the visitor's visa. The appointment calendar had been closed for months, and they now had availability for late November—right when I was also planning to visit New York to launch *Terre.*

Gaba's shift ended after the formal Parisian lunchtime was over (most restaurants in France serve meals until 2:30 p.m., sharp, and then they shut their doors until dinner). That meant we were resigned to either cooking at her place or meeting at one of the *service continu* bistros that served mediocre versions of French classics like *croque madame* and watery *bière blonde* on tap. This we did often, sharing a platter of *escargots* while scribbling business plans on the paper table covering. We postulated various names for our forthcoming wine bar (Tuesday Addams, the Brooklyn) as well as "concepts" (spritz-focused drinks menu, all-Italian natural wine, small plates). Then, we'd walk the streets of the Right Bank, soaking in whatever we could of the autumn sunshine, running various errands—dropping off shoes at the cobbler, picking up cheese for some dinner party we'd been invited to—and smoking (Lucky Strikes for Gaba, rollies for me). After Gaba's dinner shift, we'd meet somewhere for cocktails or she'd return to the apartment with a half-finished bottle of wine from work.

One day toward the middle of October, we were sitting in the apartment, drinking Beaujolais. Gaba was ranting about rude customers who had demanded, without calling ahead to warn, that the chef prepare them a vegetarian tasting menu, something very *faux pas* in Paris.

I had news to share with Gaba, something I'd been holding onto for a week. Wildman and I had just confirmed it, a few days earlier. I was excited, yet apprehensive at the same time—for several reasons.

"When we open Tuesday Addams, let's put a sign up saying, *pas de vegetarians,*" Gaba half-joked, pulling on her Lucky Strike.

I laughed. "We definitely will. But speaking of our wine bar, I have some news! I got an invite to Australia."

Thanks to a bit of emailing on Wildman's part, I'd received an offer from the Australian national wine association: they would fly me over there to attend Rootstock, a big annual tasting of natural and artisanal wine held in Sydney. They would also arrange for me to visit winemakers in a few other parts of the country, including the Adelaide Hills, conveniently where Basket Range was located.

Gaba put out her cigarette. "That's a pretty good deal," she said. She hid her eyes below her overgrown fringe, but her gaze seemed to evade mine.

"I think I'll go after my trip to New York."

I maintained a carefully measured tone. After all, Gaba and I had been planning to reunite in Paris, and here she was, only two months in, and I was already making my exit—temporarily, of course. But still, it was a change in our plans.

"For how long?" Gaba asked, lighting another cigarette.

"Maybe five weeks—it'll be summer down there! There's a big wine event, and then I'll spend a few weeks with Wildman," I said. "I'm thinking to head back to New York after the trip to finalize my French visa, and then come back here just in time for la Dive in February."

"Well, whenever you come back, you can stay with me," said Gaba. "And I'll just keep looking for places to rent in the meantime. Ideally someplace that comes with a liquor license. It's so expensive to get one here."

We finished the bottle and decided to call it a night. Just before falling asleep on Gaba's couch, I flicked through all the photos and videos Wildman had sent me showing his farm and friends and restaurant in South Australia—it was an enticing life, so different from the one I'd been living—healthier, more wholesome—and I was looking forward to seeing it in person.

Come November, I had to face up to the fact that my time in Paris was winding down. The plan was to travel first to the US to launch the magazine in New York alongside the second annual RAW WINE fair. After RAW, I'd visit my family in the Washington, DC, area and then head to Australia.

Wildman called and said he couldn't wait to see me—and he also apologized for the occasional bouts of neediness and separation anxiety he'd displayed before. "I'm learning to enjoy being alone," he promised.

His confession had an impact on me. During those weeks we'd traveled together in Europe, Wildman had shown that he was not only funny and intoxicating with his unpredictable energy, but also kind and caring. However, I'd been effectively single for years and couldn't imagine that I'd be able to commit to the stable relationship he sought. And whatever he said on the phone, I knew he was looking for something serious and lasting. Was I?

I felt torn about going to Australia, especially as it could potentially mean abandoning the wine bar plans with Gaba. She and I had become like sisters. Both of us being new in Paris, we shared the experience of living in and leaving New York as well as the emotional highs and lows of trying to fit into a city that can seem very tightly woven socially.

Even with the good fortune of an all-expenses-paid trip to Australia and knowing that I could further explore my romance with Wildman, it was difficult to think about leaving Paris. My French was improving by the day. I was starting to feel like a regular at the natural wine bars of the 11th arrondissement. So, I continued applying for the elusive French visitor's visa.

I stopped by Le Verre Volé when I knew my old friend Louis would be working. He and I hadn't hung out much since I'd moved to Paris, but he was kind and reliable. "Can you help me with paperwork?" I asked Louis. I needed a letter from someone who knew me professionally—he'd watched me grow my writing career over the years through multiple trips to France—as well as something saying I had a place to live, and he agreed to help with both. I also asked to borrow several thousand dollars from one of my older sisters to pad my bank account. I needed to appear financially stable, to convince the French government that I would not try to work illegally.

I stashed a suitcase of things in Gaba's apartment, promising I'd be back a few months later to collect them.

"I'll see you soon," she said, making light of things.

"Yes, and good luck with that guy, what's his name?"

Gaba had just met an American guy who was in Paris for an advertising job, and they were dating intensely. His name was Dan and she seemed to really like him. I was glad to be leaving her with some romantic prospects. Paris was working out for Gaba, far more than it had for me.

As I boarded a flight to return to the United States for the first time in over five months, I felt a thrill—because I would be returning to New York with a new publication under my belt. It was small and indie, but it was something I had helped to create. I hadn't even seen it yet, in the flesh.

In New York, I held the magazine in my hands for the first time and felt like a proud editor-mother—it looked great. There was my feature on Julien Guillot, and the other articles I'd worked so hard to edit, over the course of months. At the natural wine event, RAW WINE fair, which was being held for the first time in Brooklyn, *Terre* had its own table. I stood behind it, sipping on wine that friends dropped off to me, while people I'd known for years approached to offer congratulations and purchase a copy. Maybe Evan was there; I didn't notice or mind his presence. I felt accomplished, a bigger person. It had been less than half a year since I had left New York, but I was moving on. No longer was I a mere freelancer, dependent on the whims of mainstream media and its disdain for natural wine journalism. I had set out on my own.

– Seven –

Welcome to Oz

I was nervous about the long flight to Sydney. The plane would depart from New York and make one stop in Los Angeles, where I'd change to the fourteen-hour flight over the Pacific. Would I arrive stooped over by exhaustion and jet lagged, my skin puckered with dehydration, unable to get through a full day awake? Australia was much farther away from the States than Europe—the time difference put it a day ahead. At least I would benefit from the current hemispheric positioning by enjoying the onset of summertime, rather than sinking into the darkness of winter directly after Thanksgiving, as I had done my entire Northern life.

It had been three months since Wildman and I had parted ways in Barcelona. Our communications since then had been alternately stressful and confusing, and romantic and enticing. How would it be to see him, now, in his country? What would it be like to stay in his home? Meet his friends, people I'd waved at through my phone screen—the couple of Manon Farm; Wildman's partners in the restaurant, Aaron and Jasper? These anxieties crept up as the day of my flight approached. But I tried to rally myself by stoking a long-held feeling that there is nothing more exciting than glimpsing a new city or country for the very first time.

My friend Nicola, who'd lent me a couch for a few weeks in Paris, had sent me a link to a guide to "Aussie slang." I studied it as my flight traveled westward, nibbling on a semi-revolting chicken and rice dish and trying to tune out the man snoring beside me. *Arvo* meant afternoon, a *bogan* was an Aussie redneck, *having a go* was trying something out, and I might grab drinks out of an *esky* (cooler) if I was at a *barbie* (barbeque). I couldn't really see any of those words rolling off my tongue during these next five weeks Down Under, but at least I knew, having watched *Crocodile Dundee* like most other Americans of my generation, to expect people to call each other "mate." And I was prepared to say "no worries" in reply to just about everything, thanks to my time spent around Wildman and Tom.

The Australian national wine association, which had organized part of my trip, assured me that my hotel would be near the "laneways" and within the "CBD." Were laneways like freeways? (The opposite: they are pedestrian walkways, often adorned with street art.) Was "CBD" a zone where I could easily purchase medicinal cannabis? (Definitely not: that stands for central business district.)

It was morning when the plane touched down in Sydney. I rubbed my stiff neck and washed my face in the bathroom before collecting my bags. My heart raced as I exited the baggage claim. Wildman had asked for my flight information, and I suspected there was a reason for that. As I peered out into the crowd surrounding the exit, there he was, in a T-shirt and jeans, wearing the same backpack he'd brought over to Europe. Wildman must have arrived from Adelaide only a few hours earlier. His tousled hair, and the goofy grin that came over his face when he noticed me, reminded me of our nights in London and Paris and Slovenia and Spain, and I felt an undeniable surge throughout my body.

"I thought I'd welcome you in person," he said. There he was, in the flesh—after months of phone calls and FaceTime sessions from halfway across the world. We embraced, and the dank smell of his perspiration was like the finest cologne to me.

"Hang on tight," Wildman said as he pulled away from the curb of my hotel.

I clutched him tightly around the waist, leaning into his back. I had to give him credit—renting a scooter to show me around Sydney was guaranteed to make me forget all of my anxiety and jet lag. The sun kissed my bare arms as we rode through some of Sydney's shopping districts, past the waterfront, where Wildman pointed out the iconic, gleaming white Opera House, which stood like a majestic swan towering over the harbor. Its modernism was a reminder of how young the nation of Australia was—not even 250 years ago, it was settled as a British penal colony. Then we rode into a neighborhood of tree-lined streets and parked the bike outside a restaurant.

"OK, so this might sort of beat Paris on a bicycle," I told Wildman as we removed our helmets.

He nodded. "I'm not much for cities these days, but I do like Sydney." He'd grown up in the suburbs there after his family had migrated from South Africa when he was fourteen.

The owner of the restaurant immediately recognized Wildman when we entered. She effused praise about his wines as she sat us and asked when the latest vintage would be available. As we lunched on freshly handmade pappardelle, Wildman explained that Australia has a large population of Italian immigrants, most of whom had arrived following World War II—and therefore, plenty of Italian eateries.

He paused with his fork lifted, and I caught his eye. We looked at each other. We'd met in Georgia, hooked up in Paris, London, Slovenia, and Spain, and here we were now, in Sydney. We both knew my arrival in Australia meant that we'd embarked on something serious. This wasn't a fling on the road. This was, potentially, true intimacy.

He broke the silence first. "And how was your magazine launch?"

I described the enthusiasm shown by hundreds of people who bought a copy of the publication at RAW WINE fair and our packed launch party at a Brooklyn wine bar. I'd also hosted a ticketed dinner with a chef friend, and we'd poured some Lucy Margaux that Wildman had sent specially for the event. Wildman held my hand under the table and listened intently as I recounted these triumphs.

"How are you going with the jet lag?" he asked me as we sipped on espressos.

"This helps," I said. "I'm definitely struggling, though."

We rode the scooter toward the coast. It struck me how low the buildings were in most of the neighborhoods we passed through, and how many trees lined the streets—Sydney was far less developed or dense than New York. The salty air pushed back against my fatigue, and I wrapped my arms tightly around Wildman's muscular torso.

"Well, this is nice," I understated, removing my helmet as we arrived at Bronte Beach. A handful of very attractive modern apartment buildings and homes, surrounded by lush greenery, were built into a hillside encasing the sandy half-moon around the bright blue ocean. We stripped down to our bathing suits, and Wildman laid out a towel he'd stowed away in his bag. There were people playing volleyball barefoot on the sand, and plenty of dog walkers in exercise gear.

"Something to drink, my dear?" Wildman opened his trusty leather hip bag and pulled out a small cooler pouch, from which he withdrew a bottle of Champagne—but not just any Champagne. This was Vouette et Sorbée's biodynamic, amphora-fermented, single-vineyard sparkling wine. He popped the cork on one of my favorite wines in the world and filled two plastic cups—never mind the lack of glassware, the setting made up for it. Forty-eight hours ago I'd been wearing sweaters and reading old copies of the *New Yorker* on my mother's couch in suburban DC. A long flight later, I was soaking up the sun, drinking fabulous wine with a handsome man on a sandy beach, with just the right breeze blowing in my hair and stunning cliffs in view.

At Rootstock, I was pleasantly surprised to find that I was considered a special guest: the cofounders, Giorgio de Maria and Mike Bennie, had set up a table designated for selling my magazine, and I presented a talk on writing about natural wine. Throughout the tasting, I was introduced to dozens of Wildman's friends, generally chefs, sommeliers, and fellow winemakers—some of whom I recognized from photos he'd texted me over the past few months.

"I've heard so much about *Terre*," one person said to me with an enormous smile. Several people asked me to sign their copies. As they walked away with their nose in the pages, I marveled at the fact that this little publication I'd dreamt up with two friends in New York was attracting interest in Australia. All those months of hard work had brought this reward.

While Wildman poured at his table for the "punters"—meaning "consumers," as opposed to members of the wine trade—I ran around tasting everything I could: Kindeli's skin-contact Pinot Gris from New Zealand's Marlborough region; Manon Farm and Gentle Folk from the Adelaide Hills; Zimbabwe-born, Italy-based Trish Nelson's second vintage of wines made in the Lazio region north of Rome—all poured by the producers themselves. I tried a Gewürztraminer from Ochota Barrels in the Adelaide Hills and several pét-nats made of Italian or French varieties. Everything had fun labels and poetic names.

It felt really different, compared to European wine tastings—there was a greater sense of freedom. The lack of appellations in Australia meant that there were no rules for winemaking—people could blend whatever they wanted, grow any grape variety they believed would prosper. This was also the case in the US, or South Africa, or Argentina—any New World wine region. But the vibe here felt particularly "anything goes." Despite feeling jet lagged, I made my best attempt to ask producers the most important questions: how they farmed their vineyards, what their overall winemaking approach was, whether they added sulfur and how much, taking notes on everything. Occasionally, I strained to understand people through their accents.

Whenever I glanced at Wildman's table, there was a mob surrounding him, their empty, tannin-streaked glasses outstretched for a taste. Eventually, he managed to extricate himself from the crowd, pulling me aside so we could devour a dozen freshly shucked, wild-caught local oysters. They were like a soothing, cool cloth on my forehead, which was tense from travel and stimulation. Then Wildman stepped away for a few moments, and I poured his wines for a few tasters, doing my best to describe them, although, I realized, I really knew nothing about how he made his wines.

"All zero sulfur added, since 2016," I said. "This one is . . . Pinot. And that's Gamay. Hm, this is a blend of . . . I'm not sure."

There were so many—at least ten on the table—and the labels displayed no information other than the wine's made-up "fantasy name." I blushed while pouring the pét-nat featuring Wildman's depiction of me on the label, remembering him presenting that wine to me in my little Belleville apartment months earlier.

When the tasting finally ended, we stumbled out in exhaustion and caught a taxi.

"We're headed to 10 William Street, probably the best place to drink in Sydney," Wildman told me. "Gio was one of the original natural wine supporters in Sydney," he said, referring to the bar's owner. When we arrived, we met Gio: a tall man with salt-and-pepper hair and a very Italian vibe, who hugged and kissed first Wildman and then me.

"We're packed down here, but you guys can hang out on the balcony," said Gio as he led us past the bar, up a narrow staircase.

The balcony was where the staff kept their lockers. A small crowd was already there, sitting on some crates and cases of wine. Wildman ordered a magnum of wine and plates of spaghetti for everyone to share. I was grateful for the cool air—it would keep me awake.

Some of the people there had introduced themselves to me earlier that day at Rootstock, and now I tried to recall their names. One woman about my age with long brown hair with fringe sat beside me with a conspiratorial smile.

"Sarah, wasn't it?" I said, refilling her glass with the light red wine, a Gamay from Jean-François Ganevat.

I'd gotten her name right. Sarah reminded me that she lived in the Adelaide Hills and worked at a winery restaurant. "It's nice to meet you," she said, looking toward Wildman, "after hearing him talk about you nonstop . . . for months!" Her attitude was warm and comforting, and I admired her clear complexion, lacking makeup. She continued, "He's been preparing for you for days, having the guys come over and tidy the house to get it ready for you. All that, for *you*."

"Really?" I pictured Wildman dutifully mopping a floor. The thought gave me a sense of confidence, having flown across the world on journalistic pretexts, essentially to visit him.

After the whirlwind of Rootstock, Wildman headed back to Adelaide, where I'd join him soon—first, I had plans to explore Melbourne and the Yarra Valley wine region in Victoria. Before heading to the airport for the one-hour flight to Melbourne, I took a walk, exploring my surroundings alone for the first time. Around the corner from my hotel, near Sydney's waterfront, I found a coffee shop, where I enjoyed an absolutely perfect flat white, Australia's signature no-foam version of a cappuccino. Walking back to the hotel, where a cab awaited, I found a woman with long, bushy brown hair standing in front of her house, with about ten brightly colored red-and-blue parrots sitting on her shoulders. They cackled loudly. The woman poured some seeds into a bin positioned in the tree, and the eccentric sight filled me with an inexplicable joy.

Several days later, squinting into the sunlight as I exited the Adelaide airport, I moved toward the red Land Rover, dragging my suitcase along. I was eager to embrace Wildman, who stood outside his vehicle, waving, in tattered gray shorts and a T-shirt. But I was immediately distracted: there, scampering around Wildman's ankles, was a puppy.

"Hi! Oh my god, he's amazing!" I stooped to meet this tiny, beige-and-white speckled creature, only nine weeks old.

"This is Alfie," Wildman said. His teenaged daughter Lucy had named the rough coat collie puppy, Wildman explained—Lucy had always wanted a dog, but his ex-wife hadn't allowed it.

I tried not to overthink the dog's role in the family's recent changes—I wasn't quite ready for that—as I climbed into the Land Rover, which looked like it had been in a war. As Wildman got in on his side, I saw that his seat was worn down almost to the frame, in the perfect shape of his right hip. We'd always driven used cars in my family, but this seemed like a next-level extension of a vehicle's life. Wildman slapped the dashboard proudly. "Fourteen years old, drives like a beaut!" Living in New York without a car of my own for eight years, I had forgotten the pride owners can feel toward these machines.

Driving along some unremarkable, suburban, SUV-packed roads along the outskirts of town, Wildman and I held hands on top of Alfie's fluffy back, and the warm air blew in through the open windows. My eyes traveled downward and noted that his nails were partly black and misshapen, his knuckles were brawny and tough—his were hands that worked every single day.

"How was Melbourne?" Wildman asked.

I'd *loved* Melbourne. The Fitzroy and Collingwood neighborhoods, with their abundance of street art, record shops, independent book stores, and adorable eateries and coffee shops, as well as their walkability, had reminded me of Brooklyn.

"It was really great," I said. "Did you know the Builders Arms Hotel is serving your Pinot Gris rosé in a spritz?" This amused Wildman—he'd had no idea about the drink. I told him about some of the winemakers I'd visited in the Yarra Valley—notably Mac Forbes, who makes Burgundy-inspired wines from organic vineyards—and that I'd seen my first eucalyptus forest and heard the rare lyrebirds chirping in their canopies.

"There's a party at my place tonight," Wildman explained. "To inaugurate my new shed." Although we'd only been reconnected for a few days, I knew Wildman well enough by then to detect the excitement lurking underneath his measured tone.

"This shed will be a real game-changer for my winemaking," he continued. "For years, I made wine outside, *au plein air*, with just a few small work sheds for storage. I got fed up with always having to run outside butt naked whenever it rained to cover up the fermenters. So, I took out a $150,000 loan and built the shed."

I knew from experience that Wildman, whose blood apparently ran warm, preferred to sleep in the nude, and it delighted me to think of him running outside, his muscly torso and skinny legs becoming slick with rain as he hauled some sort of cover over a fermenting vat of grapes.

Through the dusty car window, I observed Adelaide, the main city of South Australia: low-rising storefronts advertising antique furniture and secondhand goods, butchers and fish shops, and various automobile supply stores and fast-food chains. Adelaide looked sleepier than Sydney or Melbourne. We emerged from the suburbanism and left behind all stoplights, on a road that ascended rapidly. Just then, Wildman and I both exhaled a sigh of relief as cool air hit the car like a wave. "You feel that?" He asked. I nodded—Wildman turned off the AC and rolled down the windows.

Now, we were in the Hills, passing through fragrant, lush forests. Wildman pulled the car into a certified organic cherry farm, where he disappeared momentarily to procure a large crate of fresh red fruit, grabbing a handful for us to eat as we drove on. I relished their sweet flesh.

"Up until the 1950s, the farmers in the Hills used to ride on horseback to Adelaide to sell their goods at market," Wildman told me. "It took days." I pictured hardened settlers on their sturdy mares with sacks of corn or tomatoes strapped on the animals' sides, coming down this very steep incline. I wondered who had been there before those settlers—the answer, of course, was that it had been Aboriginal land, that much I knew. As an American, the violent and systematic destruction of native cultures was not an unfamiliar narrative.

The Land Rover cruised along country roads—no street lighting, no commercialism in sight. The path swerved this way and that, and little Alfie tumbled from the console onto my lap. Dappled sunlight streamed through the tall, spindly eucalyptus trees. Wildman grew quiet, allowing me to take in the unkempt, powerful wilderness alongside the road. We passed a bubbling creek, and the air cooled even more. I saw big gray cows ambling around on a hillside and a grove of massive pine trees the other way, and then we turned onto a gravel road, wide enough for just one vehicle.

Large white birds with fabulous yellow mohawks swooped overhead. "Cockatoos," Wildman told me. A few red-blue-and-green parrots dove toward the car and then careened away. On one side of the gravel road, a ridge careened down into a lush, fern-covered valley. The road had some potholes, which Wildman expertly evaded. I had the feeling he could drive this route with his eyes closed.

Then we saw Wildman's vineyard, the one I'd watched him pruning from afar, in videos. He had planted it with Pinot Noir fifteen years ago, likely inspired by the multiple vintages he spent working in Oregon's Willamette Valley. It had never been watered or sprayed—the only other vineyard I'd heard of or seen like this was Ramaz Nikoladze's vineyard in Georgia.

"There's the veggie patch," said Wildman, gesturing to a large plot where rows of leafy greens were interspersed with rich brown soil. Farther up a hill, a large blue shed came into sight—it was the same color as the bright, open sky above us. Just up the hill from that shed, set amongst towering evergreens and bushy dry grasses, was a humble sandstone house with a tin roof—my home for the next three weeks.

Within my bones, all the toxic energy that had accumulated over the past years while living in New York, and the stress of hopping from one couch to another in Paris, started to dislodge. I wanted it expunged from my body, out of my spirit, and this place, with its majestic trees and impeccable air, felt like the setting I'd been looking for, even if I hadn't fully known it.

"A glass? Why would you need a glass?" Aaron was only half-joking as he slugged directly from a bottle of wine, meanwhile watching me for a reaction over the thick frame of his eyeglasses. Wildman's friend and fellow restaurateur was testing me—could I party like the Aussies? I wanted to make a good impression, the first time meeting Aaron. But I really preferred a glass.

Wildman had told me much about Aaron over the months, as well as their other business partner, Jasper, who wasn't at the party that night. The three co-owned a restaurant called the Summertown Aristologist, in a nearby town. The veggie patch I'd seen on the way in earlier that day provided ingredients. I couldn't wait to eat there.

Aaron acquiesced, handing me the stemware I'd requested, filling it with some pét-nat. "Welcome to Australia," he said with an enormous smile, pumping his fists in the air to the beats blasting from the shed and taking another swig from the bottle.

It made me a bit nervous meeting people close to Wildman. What stories about me had they already heard? What did they think of my arrival, only eight months after the departure of Wildman's ex-wife? I made eye contact with Wildman, who was using his bare hands to turn a rack of lamb ribs that was roasting on a fire pit just outside the shed. He was talking with some people and raised his glass to me, then waved that I should come over to join them.

Over the next few hours, dozens of individuals presented themselves, their names dissipating into the air, eluding my memory. There were small children, who danced wildly to the disco music that Tim, winemaker at the nearby natural wine estate Manon Farm, was playing on vinyl.

As the sun began to set over the valley, we ate skewers of lamb from Wildman's own flock, and a band began setting up. Their lead vocalist, a young woman with shiny brown hair, began crooning folky ballads and strumming an acoustic guitar, accompanied by a drum set and tambourine. Everyone quieted down and turned their focus to the performance. Wildman beckoned to me, nudging me with a hand on my waist, to follow him. We slipped away toward the very back of the shed, and he pulled me close to him. We began slow dancing.

"Nobody can see us here," he told me as he pressed his lips to mine.

When we headed back to the group I noticed Aaron sending a few winks Wildman's way. Looking behind the band, I saw very clearly, thanks to the bright shed lights, where we'd just been dancing. We had been making out in plain view, entertaining all the Basket Range community.

A few people made jokes about it to us. "I'm so embarrassed," I said to Sarah. But she only laughed and said, "Oh, it's beautiful. Enjoy it." We were under the stars, the night was warm, and bottle after bottle of wonderful, natural South Australian wine was being emptied.

Sometime after midnight, I couldn't help but be overcome by exhaustion. I stumbled up toward the house and opened the sliding door—the main entrance, which had neither a proper door nor a lock. Giorgio, the wine importer, whom I recognized from Sydney, was curled up asleep on the couch. Several cases of beer and wine were strewn about on the floor. There was a refrigerator in one corner and an industrial sink tucked away in a strange adjacent room that seemed to have once been a bathroom. I passed by the room where, Wildman had told me, Lucy occasionally stayed over when she wasn't at boarding school and glimpsed someone, I wasn't sure who, passed out on her bed. There was one room that held nothing more than a red couch, and there I found two young women curled up with Alfie, the puppy, all snoozing happily.

After a few scant hours of sleep, I woke to the sound of birds chirping—it sounded like they were gurgling water. It was birdsong like I'd never heard before. Soft yellow light streamed in through the curtainless French windows, and I sat up so I could better view the valley and hills just beyond the property's edge.

Wildman's side of the bed was empty but warm. I was about to get up to find him, when I heard a dramatic crash and bang coming from the front room: the unmistakable tenor of many wine glasses breaking. Moments later, Wildman appeared with a jar of water.

"Run into something out there?" I asked. He shrugged and gave me a peck on the lips, and handed me the water.

While he was in the shower, I went to survey the damage in the front room. I found the box of broken stemware, as well as stacks of meat-crusted dishes, a disgusting pink bathrobe, and wine bottles emptied to the dregs. The white plaster walls were peeling in the cobwebbed corners, and the only adornment on them was a large square canvas bearing an enormous flower painted with acrylics—signed by Lucy. I peered through one of the floor-to-ceiling glass doors, which allowed a nearly panoramic view of the valleys around us, toward the winery. There were at least a hundred wine bottles from the previous night, strewn across the concrete entrance to the shed. And just beside the shed, there was a car with a door open, from which I could see legs dangling—the musicians.

"Thanks for the party!" They sauntered up to the house, smiling in a way that told me they'd had quite a night. Wildman emerged wearing the same T-shirt he'd had on the night before and offered everyone coffee. He then stationed himself behind an espresso maker that had a vintage look and began grinding and packing coffee. Giorgio appeared to have vacated his spot on the couch to catch an early flight back to Sydney, but the other people who'd slept in the house awoke one by one and came into the front room. Wildman offered them all coffee. Nobody seemed to mind the mess or find it surprising. I sat down on the sofa and immediately noticed cat hair all over the cushions.

The lead vocalist of the band, who introduced herself as Naomi, sat beside me and asked me sweetly how I was finding Australia so far.

"It's gorgeous," I said. "I can't even believe where I am."

She smiled in agreement. "I know. I live in Adelaide, but I come up to the Hills whenever I can—it's really special here. And this farm, it's got an amazing energy."

Alfie stumbled by and I scooped him up, inhaling his puppy scent. A hum of morning chatter commenced in the room as people laughed over moments from the night before (thankfully, nobody mentioned my slow dance and make-out session with Wildman). Wildman handed me a flat white with a heart drawn in the frothy milk.

"What's this?" I said, pointing to the strange pink bathrobe on the sofa.

"Oh, that's Barry's love blanket," he answered. When I blinked at him, he added, "The housecat."

"Ah, right," I said, using the tips of my fingers to lift it up and deposit it on the ground by my feet.

As I sipped my milky coffee, I looked at this man I was unmistakably falling in love with. And then I let my eyes wander again inside the room, taking in the smudgy windows, the streaks of dirt and cobwebs in every corner, the random blankets and piles of paper and cases of wine strewn about—the absolute absence of order.

How was I ever going to make this place livable?

– Eight –

Dear Diary

For months, I'd been essentially living out of a suitcase, carefully keeping my presence to a minimum so as not to disturb the homes of my friends in Paris. Now I gratefully unpacked my things—clothes, my laptop and its accessories, some toiletries, and a few books—in Wildman's bedroom, which overlooked a lush valley where his flock of sheep roamed freely. With satisfaction, I hung a few items in the simple white armoire in the bedroom, which I imagined had once stored Wildman's ex-wife's clothing and had been emptied for me. Although I would only stay put for a few weeks before heading back to the US to finish my French visa proceedings, I made myself at home. I did wonder if we'd get along as well as we had in Paris. Would we bicker over things like putting the toilet seat down or doing the dishes? And I wondered, too, whether I'd meet Wildman's teenaged daughter, Lucy, for whom the farm and winery were named.

Stepping out the back door onto a patio with a concrete red floor, I joined Wildman at a small, round, cast iron table for our morning coffee.

"Today's going to be a hot one!" He proclaimed, looking out over the valley below. From the depths of the valley, eucalyptus-covered hills rose up and up toward the wide-open sky.

"I couldn't be happier," I murmured. To my delight, it turned out that the South Australian summer was completely devoid of humidity. Growing up on the East Coast, I was accustomed to stepping outside in the hot months and immediately dripping with sweat. At this extreme southern end of the world, the sun burned harshly, without the protective ozone layer filtering its rays.

"So, what are the plans for today?" I asked Wildman. I'd been there for nearly a week, and we had developed a routine of sorts—after coffee, Wildman would spend an hour or two tinkering in his shed (there was always some tractor or barrel that needed repair); meanwhile, I'd feed the puppy and catch up on emails. Then we'd reconvene for breakfast and I'd join him doing some work in the winery, such as packing up orders. If he wasn't too busy, we'd spend time walking around the farm, once sneaking into the neighbor's cherry orchard for a roaming snack. Our intimacy blossomed, and I felt less worried about the state of the house and the uncertainty of my direction.

It was summer—generally a less busy, even carefree time for winemakers. The vineyards were treated with anti-fungal sprays, as necessary, and shoot-thinned during spring. Now they were effectively left to produce grapes for about three months. Wildman had only one vineyard on his farm, and he never treated it with anything—as a sort of experiment. It wasn't very prolific, but he did manage to make a barrel each year from the fruit.

I had suggested, before arriving in Australia, that we take a trip to someplace where wine was not the main focus. Reflecting on our travels in Europe, everything we'd done had revolved around wine, and I felt it would be good for our relationship to discover new ways of enjoying each other's company. Somehow, that had morphed into an eight-day trip to Vietnam. First, we'd head to Melbourne for another big wine tasting called Soulfor, this one focused exclusively on sulfur-free wines, and then we'd catch our flight to Ho Chi Minh City.

Upon return, we'd relax and enjoy the farm and maybe take a trip to Sydney.

"I thought that tonight we could have dinner at the shop," Wildman said, passing me a cigarette he'd rolled. I had noted since arriving in Australia that he was smoking quite heavily—a bad habit he'd probably learned from me in Europe.

"The shop" was what Wildman called his restaurant, the Summertown Aristologist—in part, he used the nickname because the actual name was too long. But he also liked to downplay it, make it sound like a country café, although from what I'd seen on social media, it was an award-winning farm-to-table restaurant.

We agreed on that plan for dinner. In the afternoon, I helped Wildman hand-label his wines—it was a tedious process. All the labels were printed at the house on Japanese silk-blend "washi" paper, which was made using traditional methods. Wildman did the printing at the house, with a regular office LaserJet printer, then cut the labels with an X-ACTO knife on the kitchen table. Then, down at the small green shed overlooking the Pinot vineyard, each label was rolled through a small electric glue machine and affixed to the bottles one by one—a time-consuming job. But we played loud music and joked around, flirting as we had on the bus in Georgia.

"I can't handle the waste when it's done with the sticky backs," said Wildman, referring to the normal labels that most winemakers use. I thought back to the Loire Valley, when one day, during harvest, I'd helped Agnès Mosse to label and pack some wines for shipment. The bottles had first passed through a rolling device that slapped on front and back labels—and indeed, they had left a stream of plastic backing in their wake, none of which was recyclable, now that I considered it. Already, Wildman was making me think more deeply about natural wine, and if he was obsessive about details, at least he practiced what he preached.

The Summertown Aristologist was named thus in homage to a restaurant that had existed nearby in the 1980s, Wildman explained to me as we headed out of Basket Range, passing by apple orchards and the sandstone quarry and careening along a country road. "Aristology," apparently, was the "art or science of cooking or dining."

Wildman had studied "cookery," as they called it in Australia, back in Sydney, before eventually deciding to switch to winemaking and enrolling in oenology studies at the University of Adelaide. Although he'd chosen grapes in the end, the table was his original and perhaps strongest passion.

It was late in the day but still bright as we arrived at the restaurant. Alfie was in the car with us—he was so little, we hated to leave him at the farm. We hopped out of the Land Rover, and Aaron greeted us, patting Alfie on the head, and showed us to our outdoor table, which was surrounded by young trees and shrubs. Previously, the building had been a pizza place, Wildman explained. He'd opened it, along with Jasper and Aaron, because there was "no place to eat and drink properly" in the area.

The restaurant had large windows lined with empty bottles—a hallmark of a wine bar. I recognized many of the labels—French, Italian, and Australian natural wines.

A wooden communal table was laid out with sturdy ceramic plates, which Wildman said were hand thrown in New Zealand, glazed in hues of seafoam green and beige, as well as handsome cutlery that I'd seen in my favorite Parisian bistros. The setup was dappled with light streaming through the branches of young trees.

We went inside to greet the chefs, Oli and Bree, a young couple from Melbourne, who were standing in the open kitchen. They chatted with us from behind a bar that was covered in jars, filled with their house-made preserved and fermented veggies and sauces. Oli and Bree lived with Aaron, not far from the farm. Inside, I saw that not only was the tableware deeply considered—everything in the restaurant was crafted by hand, from the wooden stools and tables, to the artwork on the walls, which apparently was from Jasper's father's collection. Downstairs, there was a wine cellar.

"Come look," Wildman beckoned. I followed him carefully down the steps.

But it wasn't wine he wanted me to see. He pointed toward the back, where a ham leg hung in one corner, salted and aging, eventually to be served as prosciutto.

"Every month we kill a pig and make it into small goods," Wildman explained. Apparently, *small goods* was Aussie for charcuterie.

Back outside, we offered a bowl of water to Alfie, who was leashed to one of the tree trunks. Into curvy, impeccably polished stemware, Aaron poured some of Wildman's pét-nat. The handmade tumblers Wildman had gifted me were also on the table—filled with fizzy water. The meal was designed for sharing and began with small goods, with bread made from ancient grains, milled on the premises, and cultured butter. Dinner proceeded with a succession of tidy, composed vegetable dishes, including a nourishing salad of freekeh, kale, and pomegranate seeds, topped with a dollop of herbed cream. I was beyond impressed.

As the night sky finally came to hang overhead, the restaurant filled with a crowd of smartly dressed diners, and Oli brought out our main courses: sustainably sourced mulloway—a South Australian fish—and chicken raised a short drive from the farm. Aaron and the chefs explained each dish's preparation and ingredients.

"Nice 'shop' you got here," I joked to Wildman as dessert landed on our table—a slice of pie made of almond meal and dried pear slices, served with thick cream. This was a proper restaurant. But he wanted it to be viewed in a humble light. He had said many times that he believed in healthful eating and drinking as a fundamental human right.

The Aristologist, as it was called for short, was Wildman's, and Jasper and Aaron's, attempt to offer that to the Adelaide Hills—and to build community in a young wine region, where natural winemaking was only recently beginning to take hold.

Wildman had generously purchased tickets for us to go to Vietnam. It was the first time either of us had been to Southeast Asia. I was running on steam from a few freelance checks that had come through, but I planned to use my credit card to pay for our hotels.

In Ho Chi Minh City's dense urbanity, once we'd showered in our hotel, we busied ourselves tracking down all manner of street food. I learned that Wildman was an adventurous eater when he ordered a plate of chicken feet at 11 p.m.

We hadn't planned our time in Vietnam much at all—if it had been just me traveling, I would have at least planned where to go. But Wildman's spontaneity prevailed. We spent our first day in Ho Chi Minh City in cafés, sipping sweet, milky iced coffee while booking flights to other parts of the country. Before leaving the capital, we visited Independence Palace, where the North Vietnamese army crashed through the gates in 1975, ending the long, devastating civil war.

In Hoi An, a small port town and UNESCO World Heritage Site renowned for historic Asian and French colonial architecture, we rented a motorbike and proceeded to get extremely lost trying to find a beach, which when we did finally get to it, turned out to be a slim, trash-strewn piece of sand alongside terrifyingly large waves. Before heading back to town, we were desperate for something to eat, but there was very little around. Luckily, we happened upon a small food stand, where a woman made us a spectacular meal of *bánh xèo*—crispy omelet, filled with shrimp and seasoning, served in rice wraps with fresh herbs.

In Hanoi, a city whose fashion industry and vibrancy earn it comparisons to New York and Paris, we marveled at the fast-paced rhythm of life. Everywhere we went, there were so many cars, and people, and buildings packed into each block. It was overwhelming at first, but we took comfort in the rich, savory noodle soups we found on street corners in the historic old city. In Vietnam, I was uncovering two interesting facts about Wildman: one, he loved history. Every place we visited, he—like me—wanted to learn and debate its role in modern Vietnam. We made it to as many museums as possible, trying to understand whatever we could about a war that is so much a part of US history yet hardly taught in schools. The other notable fact was that Wildman was a fantastic motorbike driver. Whether we were on unpaved, gravel roads going through a tiny hamlet on our misadventure outside Hoi An, or the dizzyingly busy streets of Hanoi, I felt safe on the back of the bike, my arms wrapped around his torso.

For eight days, we didn't have a drop of natural wine. Like most couples do when traveling, we got to know each other in new ways—through discovering a new place, arguing about various little things, and finding out what we liked doing together. We flew back to Adelaide in mid-December. It would be the holidays—Christmas and New Year's—and we were looking forward to slowing down and relaxing at the farm.

Little by little, I began rearranging Wildman's house. "What if the fridge was here?" I asked, indicating an unused corner. "And maybe if we move this couch around to that side of the room . . . "

Wildman, standing at the kitchen island, grating carrots and zucchini from the veggie patch for our midmorning "rosti"—a fritter of sorts—stiffened at my suggestion. He mumbled a response about plans to tear down the prefabricated walls that enclosed the front room of the house and rebuild them with locally sourced sandstone. The back part of the house, which was older, was made of sandstone bricks—but the front room had been added on as an extension.

I wasn't interested in talking about renovations—I was concerned with making it cozy for the short term. As in now.

Wildman noted my exasperation. "When my family moved to Sydney from South Africa," he explained, "we were only allowed to bring five thousand dollars per person. It was a condition of leaving, from the apartheid government." Back then, in the late '80s, his parents had started a business renovating houses, and they built that small amount of capital up over the years to the point they could purchase, flip, and resell property. Wildman, encouraged to join the family business, owned his first house at the age of seventeen.

We sat down to eat our crispy rosti. "So, I have a particular way of doing things," he finished. Then he looked around and reconsidered. "But we could try moving that over there. And this year I think I'll start working on that room—" He gestured to the unfinished part of the house where the kitchen sink resided alongside various boxes of wine and random items. One part would be a pantry, attached to the kitchen. Another part would become a guest bedroom.

I would have found it all hard to believe except that I already had proof that Wildman was very handy with tools: he had designed and built the bathroom in the house. It had gorgeous wooden floors, white tile walls, a claw-foot bath, and nicely selected appliances. He had mentioned, however, that the bathroom had taken five years to build.

The next time we went to the grocery store together, I threw a new set of tea towels and sponges into the cart, as well as an industrial broom to sweep up the mounds of dust and dirt in the front room. It wasn't much, but every day we were cleaning more and making small changes, and the house was starting to feel a bit more hospitable, less like an extension of Wildman's workshop and increasingly a place where we could coexist as a couple.

I'd heard much about Wildman's daughter, Lucy. He occasionally texted with her or called her—he liked to check in, he explained, because apparently she didn't really like boarding school. I wasn't sure how to relate, since my entire family had gone to public schools, but Wildman explained that the local Basket Range school felt limiting for Lucy once she hit her preteen years.

I had seen photos of Lucy—a long-haired brunette with movie-star lips and a svelte figure. She'd become interested in modeling recently and was taking a course to build up a portfolio. She didn't sound like the heiress to the Lucy Margaux Farm and wine brand, despite it being named for her. Wildman didn't seem concerned.

We had returned from Vietnam to the arrival of yet another puppy—this one a feisty, black-and-tan Australian shepherd, whom we called Lulu. The two dogs were immensely cute, and we occupied entire afternoons sitting on the patio, drinking spritzes made with an organic type of Campari, watching the puppies tumble into each other, play-fighting. We were doing this when Wildman mentioned that Lucy would be coming to visit the next day, to stay for a while. It was, he told me, the start of school holidays.

Despite having been forewarned, it was something of a shock when, the next day, I'd returned from the veggie patch and was unloading bundles of curly kale and handfuls of chili peppers onto the kitchen island, and Lucy emerged from the spare bedroom. Of course, it was actually her room, although I'd never thought of it that way. Wearing a tank top and shorts, and sniffling into a tissue, she eyed me warily.

Should I hug her? Wave? I smiled grandly, but before I could say more than hello, Wildman appeared from another part of the house and formally introduced us. She hardly said a word to me, instead asking her father for some toast, which he set about making. Apparently, she had a cold, from all the stress of term finals.

Lucy disappeared back into the bedroom and mostly stayed there. For several days leading up to Christmas, I watched Lucy pad around the house in thick socks with her long, wavy brown hair in a ponytail. Every morning, she took a long shower and then made herself toast, which she brought into her room, where she remained until lunch, and then again until dinner—except for visits to the bathroom. I became used to it being occupied right when I needed it, every single time.

"Lucy? Are you taking . . . another shower?" I called through the door. But soon I gave up, not wanting to be the mean new girlfriend. Wildman defended his daughter's frequent showering, saying that it was a "teenager" thing. One morning, while I anxiously crossed and uncrossed my legs, debating whether to pee outside rather than wait another half hour for Lucy to finish showering and dry her hair, I tried to remember if I had also done this as a sixteen-year-old. Again, I couldn't relate, because I had had my own bathroom as a teenager. I knew I should feel empathy for Lucy, who was going through a hard time, but when I looked for it, I found a gaping hole. I had no idea how to befriend or even understand my boyfriend's daughter.

An hour later, I finally had my chance to shower. But when I went to turn on the water, it ran cold, with no sign of warming up.

The house's water was heated by the sun, which meant the heat was limited. I'd have to wait until the next day for a hot shower. I redressed angrily, and before leaving the bathroom, quickly swooped up the pile of makeup brushes and bottles that Lucy had strewn across the counter and stuffed them all in a drawer. Clearly, she took after her father in terms of tidiness. I picked up a few used tissues and threw them away. Then I went into the kitchen, where Lucy's three days' worth of breakfast dishes sat in the sink and washed them.

I had been enjoying the domestic bliss of the house until then. But I couldn't just spill out my frustrations to Wildman—he was sensitive about his daughter, and rightly so. The family had just been through a separation. In fact, I couldn't really vent to anyone there in the Hills—I had not a single close friend around me.

Fortunately, I had the one outlet I'd relied upon since I was a little girl and my parents had begun the fighting that built up to their separation: my journal. My diary was my ultimate safe space, where I could let go of all the things I would never say or write to people directly or in public. After Lucy had been on the farm for about a week, I sat on the patio with a glass of pét-nat in the afternoon light and wrote a long entry on my hesitations toward what I was getting into.

I'd been single, albeit reluctantly so, for several years at that point. I was accustomed to dealing only with my own emotional fluctuations, my own ineptitudes and shortcomings. I'd never had to clean up for anyone else, other than lazy roommates. Now, as much as I was coming to love Wildman, I also had to confront our differences. For one, I loved to read, while he was prevented from enjoying the written word by severe dyslexia. Upon arriving to the farm, I had gifted Wildman a copy of *A Farewell to Arms*, having learned during our trip in Slovenia that he'd never read anything by Hemingway—and now the book sat untouched on his bedside table. Wildman's chaotic approach to life seemed at odds with my love of feng shui. Then again, he was sweet, attentive, handsome, and fun to be around—and an inspiring natural winemaker. I wrote my heart out, and once the words were on the pages, I felt I could begin to think it all over.

On Christmas, Lucy headed back down the hill to be with her mother in Adelaide, and Wildman and I were once again alone in our coupledom.

Most of my Christmases until then had been spent awkwardly pretending to enjoy the holiday—which in my mostly Jewish family was hardly notable—or, on one lonely occasion, reading alone in my Brooklyn apartment, with a bottle of red wine by my side and the heater on full blast. This was bound to be a new holiday experience. On Christmas morning we picked all the kale we could stuff into a basket, then strolled along the driveway with the dogs by our side, heading to the house where Aaron and the Aristologist's chefs lived. Aaron was from Queensland, and Oli and Bree were from Melbourne but didn't feel like heading back there for the holidays. So, those of us without relatives would make the day festive together.

As we passed a eucalyptus grove, I craned my neck for koalas. We'd already had a few sightings in the area.

"Close enough for you?" Wildman asked, gesturing to a tree right by the path where we stood. There was a large, fuzzy koala, at eye level just a few feet away, munching on leaves. I could see its long, black claws from where I stood. Wild koalas, I knew, were definitely not to be touched or cuddled.

We walked on. "Do you think you'll come back for vintage?" Wildman asked the question with feigned casualness, but I could tell by the high pitch in his voice that he'd been thinking about saying this for some time.

It caught me off guard, and I wasn't sure exactly how to reply. The last time I'd spoken on the phone to Gaba, she had updated me on the guy she was dating—Dan, an American. She sounded happy and was enjoying Paris, but she also pressed me about when I'd be returning there to help her with the plans for our wine bar.

"I'm really not sure yet," I told Wildman. "I still need to see if I can get that visa for France. The application wasn't complete when I was in New York, which means I have one more appointment." Surely it sounded a little ridiculous to be going on about a visa to live in France when Wildman and I had been such a close-knit couple for the past month. Was he was wondering what all this was for, if I wasn't going to be part of his life in a more permanent way? I tried to make light of it—"And you don't want me to help with your winemaking. You should have seen me trying to help the Mosse brothers. I'm great at drinking wine, pouring it, and writing about it, but making it really isn't my forté."

We passed by the creek and started up the last hill toward Aaron's house. On our left, a cluster of big gray cows were munching on the last patches of springtime grass. "Right, of course, I would *never* want you to give up on your dream. But you could come here just for a couple of months and do the vintage," Wildman continued, convincingly. Authentically, I wondered? "You could make a barrel of wine. It would earn you a little bit of cash, to pay back some of your student loans. And then you can just go back to France, if that's what you want."

Wildman was aware that my financial situation wasn't great at the moment. I had loans from graduate school. His offer to make a barrel of wine for my own benefit was incredibly generous. But it was obvious that he wasn't just trying to help me get back on my feet—he was asking me to move in with him. And to spend considerable time together, running a household and working in the same winery. I didn't really know how much money a barrel of wine could turn up. It all sounded quite improbable. But despite my hesitations, and all the work that I'd put into my French visa, and the plans I'd been making with Gaba, Wildman was definitely swaying me.

I squeezed his hand and told him I would think about it, and he shrugged and winked. That night we feasted on oysters, shrimp, and plenty of natural wine with Aaron, others from the natural wine community, and the chefs, and laughed and danced under the bright South Australian stars until the early hours.

The day after Christmas, I took the dogs for a long walk. They scampered and tumbled ahead of me on the unpaved road, all the way to the creek, where we traversed a hillside on a well-worn footpath to the Basket Range swimming hole. I spent a blissful hour soaking my legs in the cool water and laughing at the dogs' tentative attempts to swim.

Back at the house, Wildman was sitting at the kitchen table—a butcher's block, essentially—absorbed in some paperwork.

"Hi, love! I don't know about you, but I could use a glass of cold wine."

He hardly looked up in response to my greeting, and when he did I noted strangeness in his eyes and worry in his furrowed brow. I wondered if he was upset about something related to Lucy's stay with us. Surely it was difficult for Wildman, to say goodbye to her and not know exactly when her next visit would be.

Later, we sat down to a light dinner on the patio. Wildman had prepared a few salads, dressed with fish sauce and soy sauce and sesame seeds, which he quietly brought to his mouth with chopsticks, seeming pensive. I'd never seen him so untalkative. We got into bed, and Wildman looked at his phone for an hour while I read a Zadie Smith novel. Instead of kissing me goodnight as he normally did and pulling me close, stroking my back and shoulders lovingly, Wildman turned his back to me and fell asleep promptly.

His emotional distance was a shock. It reminded me of that night in London when we'd fought at 3 a.m. about me having a conversation with a bar owner. In the morning, I found Wildman in the kitchen, making coffee. He didn't glance up when I entered.

"Babe. Is something wrong?" I asked.

He flinched visibly and slowly turned toward me.

"Am I not good enough?" He spoke with bitterness in his voice. "Because if you don't want this, if I'm not right for you, then it's better we just end this right here, and now." He paused and looked me in the eyes.

"I don't understand what you're talking about." I shook my head. "One minute, we're like a happy family"—in my mind, the dogs had become like our children—"and now you're acting so cold toward me. Where is this coming from?" We stood there, looking at each other. Then, in a moment, I knew. A cold feeling went through me as I pictured my diary, a Moleskine with a neon pink cover, sitting in plain sight on the bedside table.

I revealed my suspicion: "Did you read my journal?"

It was a strong accusation, but Wildman didn't deny it. As he tried to make excuses, I ran over the recent passages in my mind—they had not been kind to him. Aghast, I ran into the bedroom and shut the door. He stood outside, trying to continue our conversation, while I dressed and grabbed the notebook. I stormed past him, refusing to listen, and ran out of the house through the back door. I could hear

Wildman calling my name as I made my way down the driveway. But where to go? I had no car, was miles away from anything, and didn't really know anyone I could confide in.

I walked all the way down the driveway to Aaron's house, where we had spent Christmas. I found one of the chefs packing his car for a weekend away at the beach. Noting my distress, he told me the door was unlocked and to make myself at home, then drove off. I was all alone to fume.

The patio overlooked another gorgeous Basket Range scene—avocado and grapefruit trees, the neighbor's field of fiery red flowers. For an hour, I alternately cried and wrote my feelings. Everything came out: that this relationship, rich with passion though it was, also felt pointless because it was on the wrong side of the world. That my being in Australia made no sense. That what I needed to do was return to Paris and carry on trying to write about natural wine and, once I got my visa, continue working with Gaba to open the bar. I wanted to believe in our wine bar project, even if it was only a pipe dream scrawled on the paper menus of all-day bistros. Believing in a wine bar felt easier than believing in love. For me, it made more sense. Especially in that moment, when Wildman had betrayed my trust.

This was cathartic. I had not had a morning to myself in weeks. In Paris and New York, I'd passed hours alone, reading and scribbling in my journal in cafés. I felt smothered by Wildman—being in a relationship was an emotional burden that I was not used to bearing. How could I get him to understand that I needed space? That to thrive, emotionally and intellectually, I had to be free to work through whatever I was experiencing, in my journal or with friends, and that I couldn't love him without being able to do this?

As I reflected on the past few weeks, I also sympathized with Wildman. I was flip-flopping on him, one day showing him affection and potential commitment, the next mentioning my desire to return to

France, my uncertainty about continuing our relationship. He must have felt like he was treading on unsteady ground. Less than a year after his wife had left, this was probably not the best way to feel.

I went inside and located the house's communal beer stash and some olives and cheese in the fridge. Back on the patio, I cut into my breakfast, finally relaxing—but then I heard the unmistakable whir of Wildman's Land Rover. He walked up with a puppy in each arm. Angrily, I rolled my eyes. Of course he'd guessed where I was. And he'd decided he would win me back with puppies. For five minutes he stood there, trying to talk with me. He left, having been unable to break my silence, but the puppies stayed. I watched Alfie and Lulu play-fighting for another hour, before slowly beginning the hike back to the house. He hadn't really apologized, and I hadn't forgiven him. But I wasn't going to just stay at Aaron's house all night. We needed to deal with this breach of trust.

Wildman looked incredibly glad to see me, and also a bit embarrassed. He opened up a bottle of his Pinot Noir and poured me a glass, and we stood there, sipping, and eventually started to talk. I told him that I needed space to adjust to this new relationship, and that it all felt rushed. He admitted he was having some anxiety, which I put into therapist's terms—"I think you mean, 'attachment and abandonment issues.'" He shrugged at my diagnosis but swore to never again read my diary or disrespect my privacy in any way.

We sat on the couch, dirty as it was with dog hair. A fiery red sun was setting over the hills, and I watched it while also seeing, out of the corner of my eye, several large black flies swarming in the corners of the kitchen. We'd need to put up some fly tape. We needed to wash the couch covers. We couldn't just drink wine and pretend that everything was OK; we had to make this house fresh and inviting or else our nascent romance would get caught in a cobweb and die tragically.

Wildman cleared his throat.

"In South Africa, as a child, I ran around barefoot, fishing and swimming naked, and I skipped school whenever I could," he began. He mentioned again the severe dyslexia that prevented him from becoming a reader and disciplined writer. I thought of the poems he'd drafted for me when we were in Europe, written in careful cursive on notebook paper. Those poems were always full of spelling and punctuation errors. But they were also full of kindness and sincerity.

"I was raised by two professional journalists," I replied, "in a house with floor-to-ceiling bookshelves. I don't know anything about farming. Or winemaking." I wanted to add: I also don't know how to live in a house without hot water, or how to be a stepmother, or how to be a wife.

We couldn't reconcile some of our differences. But natural wine and the ethics behind it was firm ground upon which we both stood. We would try to move forward with this bond, and maybe it would suffice. Wildman said, again, that he wanted me to come back for vintage. I said, again, that I would think about it.

In my diary, I'd written about Wildman's faults, his lack of intellectualism. But had I recognized my own shortcomings? Had I accounted for the ways I was limiting myself, was afraid to test my own boundaries?

After our conversation I decided to give Wildman another chance. But as December neared its end, I was increasingly skeptical about the idea that I'd be returning to Australia for vintage. I'd just launched a magazine and wanted to concentrate on the second issue. I wanted to return to France and visit some of the regions I hadn't yet seen, like Beaujolais. I wanted to write a book, about natural wine, and where better to do that from than Paris? And who knew, maybe Gaba would secure funding, and our wine bar, Tuesday Addams or whatever we called it, would proceed.

Wildman and I drove to a quiet beach south of Adelaide, with two new sets of snorkels and flippers. We walked past the small crowd suntanning on the sand and eventually came to a secluded cove where the water was pure turquoise. After I managed to get my mask on properly, I grabbed Wildman's hand, and we stumbled over the rocks until the water came up to our shoulders. We dipped underwater and approached the corals. They were full of sea anemones, urchins, various small fish, and even an octopus. Its gray-and-white tentacles writhed along with the rhythms of the tide, and I treaded water, mesmerized. Back on the shore, we drank a bottle of wine out of plastic tumblers.

We spent a festive New Year's Eve at the Aristologist. The restaurant filled with revelers from the Hills and Adelaide and people from overseas who'd come for harvest internships. We ate small goods and locally made cheeses with the Aristologist house bread, and it was one of the most enjoyable ways I'd called in the year's end in recent memory.

A few days later, I once again boarded a twenty-four-hour flight, now returning to New York, where I had two goals: one, to move my very last belongings down south to my mother's house in Virginia, with the help of a friend; and two, to finalize my French visa. On the plane, I thought about my brief experience in Basket Range: the massive pine and eucalypt trees, the satisfaction of caring for pets, the stress but also magic of cohabiting with a loving partner. Not to mention the thriving natural wine scene I'd discovered in Australia. Wildman's offer, for me to return for vintage, stood. I'd told him I would make my decision soon, and he seemed confident that the answer would be yes. I wasn't quite sure. Winemaking and love seemed equally beyond my capabilities.

It was blizzarding as I landed in New York—and there was bad news. My partners from *Terre* were stranded, one in Miami and one in upstate New York. Our plan had been to spend a day or two working on the magazine, planning future issues. I dragged my suitcase through the snowy streets of Williamsburg, going right past my old stomping grounds, Uva Wines, to a last-minute couch-crashing arrangement.

On my way to the French embassy in the morning, I stopped into Toby's Estate for coffee. It had been one of my regular writing spots back when I lived in Brooklyn, and I surveyed the scene of laptops, hot chocolates, and thick parkas. At one table near me, one woman and two guys were staring intently at the woman's phone screen, evaluating whether she should reply to someone's conversation starter on a dating app. They scrutinized what the candidate was wearing in his profile photos. I smiled, remembering the dozens of terrible online dates I'd been on in my Brooklyn days. I wondered if I would ever have casual sex with a random stranger again, or if those impulsive days were a thing of my past now.

My phone rang, and when I saw it was Gaba, I answered immediately.

"Hi, uh, sorry, uh, are you busy?" Something was off in Gaba's voice.

We chatted for a moment about my flight to New York and the blizzard. Then I heard her take a deep breath.

"I have some news . . . I'm, well, I'm pregnant."

A few things flashed in my mind. One was an image of Gaba, wearing dark Wayfarer sunglasses and a carefully knotted *foulard*, pushing a stroller along the sidewalks of the Marais. Another was of the two of us, standing behind a bar, laughing, loud music playing, pouring glasses of natural wine for a crowd of good-looking Parisian men.

My friend had only been dating this American guy, Dan, for a few months. I hadn't met him, but Gaba seemed to like him. And now she was going to be a mother. She had just started working at a really hot bistro, a job I knew she was excited for. That probably wouldn't last now.

I'd never heard Gaba mentioning her desire to be a mother before, but she explained to me now that a doctor had once told her she couldn't get pregnant. And yet nature had proven him wrong. Or rather, the one time that Dan had come over to fix Gaba's flickering light bulb, and they'd ordered pizza and opened a bottle of wine, and someone had been too lazy to run out and get condoms, had proven the doctor wrong. Actually, Gaba sounded really happy. She was one of the most interesting and intelligent people I knew, and maybe she had a bit of a wild streak, but I could see her becoming a wonderful mother.

But my second vision, of us behind the bar of our own café, the one we'd dreamed of opening together in Paris—would that ever happen? The question hung in my mind as I walked into the French embassy uptown, handed over the last of my paperwork, and received a visa stamped into my passport. I was now legally permitted to live in France for one year.

It was time to leave New York. My friend Megan, who had worked for a natural wine importer, drove from Philadelphia to Brooklyn to pick me up and take me, and my belongings, to Virginia.

"You're the best," I told her. She was doing me an enormous favor. In only six months, it seemed I'd lost touch with so many of my New York friends, something I hadn't really anticipated when I made my plans to move to Paris. But Megan and I had shared many bottles of natural wine together over the course of our friendship. We had supported each other through breakups and professional crises. It seemed our friendship was destined to survive my move across the pond, or across the planet.

As we made our way out of the city in her reliable old sedan, I updated Megan on my romance with Wildman and our adventures in Melbourne, Sydney, and Vietnam. I described the farm, his restaurant, the party he'd thrown to celebrate his new shed, the puppies, our morning coffees overlooking the valley, the anxiety I felt over meeting Lucy.

"He sounds great, Rachel," said Megan. "I'm really happy for you. Are you going to go back, for vintage?"

I was finding it difficult to answer her. I had finally procured a visa to stay for one year in France. But ultimately, what did I want? What was best for me? To return to Paris and be, once again, untethered, dating, scraping together a living from freelance articles while preparing the next issue of *Terre*? What were the chances, really, of Gaba and me getting our shit together anytime soon and making our wine bar a reality—especially now that her life was changing in the most dramatic and permanent way possible?

"I'm not sure," I finally answered Megan. "He's so loving. He's a genius and has created something really amazing in South Australia. But he's also wounded, and it makes him needy and jealous. We fight constantly, and I don't know whether to trust him, and then we make love or have a special meal and I feel like he's my soul mate." Exasperated, I looked to Megan to help me make sense of this situation. I'd meant to move to Paris. How had my life been derailed by one man?

She shrugged, staring out at the freeway. "Sounds like love to me."

I blinked. Was this what love was actually like?

At my mother's house in Virginia, where I'd spent my teenage years, a stone's throw from my public high school and six miles from the White House, I unloaded all my boxes. After I took her out to dinner in thanks, Megan made her way back to Philly. Back in my old bedroom, I added my anthropology texts from grad school and stacks of novels to a tall shelf, which was heavy with various belongings from my teenage years. There, I found an old book that I'd loved as a child—a collection called *D'Aulaires' Book of Greek Myths*. Flipping through, I came to one of my favorites, the story of Persephone. Her mother, Demeter, was the goddess of agriculture.

One day, Persephone was out walking, and Hades, the god of the underworld, glimpsed her and fell in love. He pulled her down into his caves. When Demeter realized that her beloved daughter had been abducted, she fell into despair and refused to let the sun shine. The whole world plummeted into eternal winter.

Zeus, who famously spawned dozens of earthlings and gods in his career, was Persephone's father, and he pleaded with Demeter to be reasonable, as the never-ending winter was proving unbearable and famine was imminent. Demeter conceded by teaching farmers how to sow wheat, which could be stored—and so people survived, but just barely. Finally, an agreement was reached with Hades that Persephone could come up to the above-ground world for half the year. During that time, Demeter would allow things to bloom and grow. The rest of the year, Persephone would retreat to Hades's lair, and winter would preside.

The dark comedy of the myth spoke to me, drawing an emotional connection between this point in my life, which was feeling increasingly like a critical moment, and my childhood years, when I'd daydreamed and written stories and known nothing about financial debt, career struggles, or the complicated desires of wanting to have a partner and start a family.

Something about being in my mother's house, where I'd longed, year after year, to escape my suburban life and never return, brought forth a decisiveness in me. I called Wildman. It was probably 1 a.m. in Australia, but I had news to tell him.

– Nine –

La Dive Part Deux

I arrived in France in mid-January to meet Wildman and Gaba and head to the Loire Valley for another session of la Dive Bouteille. My new, hard-earned one-year *carte de sejour* burned a hole in my passport.

Upon landing at Charles de Gaulle Airport, I took public transit straight to my very favorite "hotel" in the city: Gaba's couch. It was a different couch, in a different place, but it still promised the comfort of an old friend from New York, someone who understood what I was going through—and she, too, was going through some significant life changes.

Gaba had moved into a studio apartment on Avenue Parmentier, within walking distance of everything one could need (pharmacy, Métro, natural wine bar). She had a bartending job nearby. By day, she mostly slept off her first-trimester morning sickness, while I wandered Paris in the dreary, cold midwinter rains. I stayed warm by remembering sunny days on the farm in Australia. And yet, as I nestled into a café crème on the terrace of a random café after walking through the Brancusi studio, where the sculptor had crafted his work while living in his adopted city, I had to admit to myself: I was still very much in love with Paris. But I was starting to realize that being in love with a city wasn't the same as loving an actual person.

One night when we were out having a glass of wine, I told Gaba that my plans had changed—I would be flying back to Adelaide after the salons. She shrugged and acted cool about it, saying that it would be great for me to get more winemaking experience, and reminded me that I could return and stay with her anytime.

A week later, Wildman arrived and we took a train to the Jura. We then embarked upon a week of touring France before eventually winding eastward to the Loire. By the time we arrived at the salons, we were exhilarated from visiting our winemaking heroes—Alice Bouvot of Domaine de l'Octavin, Kenjiro Kagami of Domaine des Miroirs, and Aurelien Lefort were highlights—and our livers were already trashed. But the drinking had hardly even begun.

It was my second time attending la Dive Bouteille, and for Wildman and me, our first as a couple. We dove into the bacchanalia: oysters and sips of Vincent Laval's Champagne for breakfast. Hours of standing around on our feet, which were snugly clad in our winemaking boots. Bump-ins with the New York and Paris and London natural wine cognoscenti. Kisses every five minutes, on both cheeks. Sandwiches at 3 p.m., espressos at 5, then out of the caves, to the parties. Dancing after dark in a crowded winery, eating risotto prepared by the chef-owner of Septime, a Paris "neo-bistro," as farm-to-table restaurants were called, glass of Gamay in hand. Every bite of raw-milk cheese a reawakening of my body's cells, crooning Edith Piaf tunes in the street with burly French winemakers, into the early hours, until we collapsed, exhausted. Rising late in the day in the most picturesque country *gîte*, a French guest home, surrounded by snow falling.

Back in Paris, we had dinner at our favorite bistro, Les Arlots. Their signature dish, sausage and mashed potatoes with gravy, warmed our insides, and we finished off the meal with shots of Chartreuse. Wildman headed off on the plane, back to Adelaide, where I would eventually join him.

On my last night in France, Gaba was at work, and I planned to go to sleep early, anticipating the long flight to the Southern Hemisphere. I lay on the couch, sipping tea to ease a sore throat, and replayed scenes from the past few weeks in my mind: our dinner of baked potatoes and melted Mont d'Or cheese at Alice Bouvot's house in Arbois, the incredible array of wines for tasting in those caves in Saumur, holding Wildman close to me in the coziness of our bed in the middle of rural France after a long night of partying. I was just drifting off to sleep when the door opened: it was Gaba, who'd returned with a bottle of sparkling red wine from La Sorga, a winery in the Languedoc region of Southern France, known for their unpredictable and imperfect, but delicious, bottlings.

I couldn't refuse one last glass with my best friend. Gaba and I were in our most honest element in that moment, and the truth was, we weren't perfect humans. We jumped into things without thinking them through carefully. We let life take us on its ride. For some reason, it was who we were.

"Babe," I finally yawned, "I gotta try to get some sleep." The couch beckoned.

"Of course," she said, beginning to make her way toward the bedroom. I began fluffing up the pillow, eager to close my eyes for a few short hours.

Then Gaba was beside me, her hand on my forearm. It squeezed, tightly. I looked at her and saw a stricken expression.

"Please don't go, Rachel," she said. "Please don't leave me."

I didn't know what to say. Gaba backed off, seeming embarrassed at her impulsive action. I rubbed her shoulder. "I'm not leaving you—

I'll be back, of course. I wouldn't miss you having your baby." She slinked off to bed, saying to come give her a hug before I left to the airport.

What *would* become of Gaba's freedom, I wondered, after her child was born? She and Dan hardly knew each other, although they seemed very much in love, or at least the early stages of it. She'd been out with him that day, looking at apartments where they'd soon move in together to raise their baby. What of our dreams to open a wine bar? More importantly, what would become of our friendship? I was leaving Gaba at her time of need, and it didn't feel right.

Broken glass: that's what I thought of, bizarrely, when I heard the operatic birdsong of magpies. I slowly opened my eyes to see the pink sky transitioning from the cool, gentle cover of night to the warmth of day.

Wildman's side of the bed still radiated heat from his presence. From the kitchen, I heard the sound of the coffee maker. I also heard the pitter-patter of small, canine footsteps. I crept over to the bedroom door and swung it open, then leapt back into bed. Within moments I had the company of two furry, panting bodies beside me. Alfie and Lulu scampered around, competing for my cuddles.

I stroked their soft bodies, savoring the early hours' calm and quiet. I heard Wildman's footsteps in the hallway and smelled the aroma of coffee. He came in, wearing only boxers. "Morning, beautiful!" He placed a steaming mug on the bedside table and began dressing. "Just going to make a run to the airport to get Alex," he said.

I remembered—the young intern from Germany was arriving today. And soon, Raphael would also be here, from London. Sev, a French woman who worked at the New York wine bar the Ten Bells, was in Australia, ready to come on board with team Lucy Margaux for vintage 2018. All three of them would be staying with Wildman and me, crowding in—in fact, the number of people surpassed the number of bedrooms in the house, it occurred to me. It appeared that "my" office, the one room in the house that I controlled, where I worked at a desk, would be a bedroom for Raphael, who would sleep on the pull-out sofa.

Our romantic household was about to get very cramped.

– Ten –

The Green Gamay

"Good morning!" Sev's French accent shone through even after her years of living in New York. It was very early, but she was enthusiastically scrambling a pan full of eggs. Beside her, Alex was slicing bananas into a dish. I blinked and rubbed my face. Our quiet mornings had become very busy now that we had a house full of interns.

I saw Wildman standing outside in his boots and boxers, shirtless, smoking a hand-rolled cigarette and sipping from a mug. As I reached for the French press, Raphael picked it up and poured a cup, which he handed to me.

"Voilà," he said, smiling. The toaster popped, and he turned to inspect the bread. "Perfect timing!"

I placed the coffee on the table, whose jagged surface had been used many times to cut labels for Wildman's wines, and sat.

Soon the table was set, a pan of eggs was placed in the middle with sliced fruit and buttered toast, and everyone dug in. Wildman came inside and grabbed a piece of toast before marching through the house as if looking for something. I ate a few bites, then got up to feed the puppies, who were happily digging holes amongst the rampant weeds and fennel bushes just outside the sliding doors.

By 7 a.m., the dishes were piled in the sink, and we were piled into Wildman's Land Rover. The car filled with the dual scents of sunblock and cigarette smoke as we drove toward the town of Hahndorf. I selected moody electronic music, hoping for one thing that it would drown out the sound of the trailer rumbling along behind us. It worked—within minutes, the crew was asleep.

Wildman nudged me and pointed to what he saw in the rearview mirror—Sev's head was resting on Alex's shoulder. "Intern romance," he suggested.

It was the first year that Wildman had so many interns, he'd explained to me. As a result, he hadn't really given thought to *how* everyone would coexist in the little house. The tight quarters had prompted a kind of redecorating of the front room. Over the weekend, when we had a day off from picking, Sev and Raphael and Alex and I worked in concert to rearrange the furniture. Now, the sink, stove, and refrigerator were together, near the rickety kitchen island. There were still various pots and pans scattered atop cases of wine. It wasn't exactly the cover of *Good Housekeeping*, but it was much more suitable than before. We also moved around the couches, swept the floors, and even dusted off some of the cobwebs in the corners of the ceiling-high windows.

I rested my hand on Wildman's knee. He seemed to be concentrating as we drove—thinking about the grapes we were headed to pick and what he might do with them once we brought them back to the winery. That shed, which months earlier had sat new and empty, was now slowly filling with barrels, stainless-steel tanks, and ceramic eggs. Over the past week, we'd picked Pinot Noir, Pinot Blanc, and Chardonnay—mostly for sparkling wines, which are harvested early for acidity. The large electric press had already been used a few times.

Hahndorf, one of the first towns in the Adelaide Hills to be colonized—by Germans, as the name indicated—was a solid forty-minute drive from the farm. On the way, Wildman explained that he'd begun working with this vineyard about seven or eight years ago. At that time the owner, Rob, had been struggling to find a buyer for the fruit. Wildman loved the schist soils, which he said brought a lean, tight character to the wines, and the fact that it hadn't been heavily chemically treated. But he wanted to expand on the varieties planted there—at the time, Rob's vineyard was primarily Sauvignon Blanc, Chardonnay, Pinot Blanc, Merlot, and Pinot Gris. After a few years of making wine from that fruit, Wildman suggested they diversify, and Rob agreed. Wildman ordered vinestock and grafted on varieties that he found more exciting: Cabernet Franc, Sangiovese, and something quite exotic in Australia—Gamay.

The day before, as we finished up a big pick of Pinot Blanc and Pinot Gris, Wildman measured the Gamay's ripeness with the refractometer. Back at the winery, as he supervised Alex on the forklift, putting the Pinot Blanc through the press and transferring it to the ceramic eggs, then destemming the Gris, which would soak on its skins for a few days, he informed us that the Gamay was ready to pick. Processing went late into the night—we finally had dinner around 11 p.m. But here we were, rubbing our eyes, buckets and secateurs in hand, approaching the ten rows of Gamay.

"Good morning!" There were the pickers—a crew from Thailand and Laos. Their ringleader was Sam, who knew the vineyards like the back of his hand, as he'd been working for Wildman and Rob over many years. The pickers wore cloth over their faces and long sleeves to protect their skin. They snipped grapes incredibly fast, so rather than picking alongside them, Raph, Alex, Sev, and I had taken to running their full buckets to the bins and sorting through the clusters to remove leaves or unsightly grapes.

As our crew started collectively moving down the first Gamay row, Wildman gave me a peck on the lips. "I'm off to clean the press—save me some smoko?" Then he slipped away toward his car. *Smoko* was the way Aussies referred to the midday break. It had originated as a cigarette break, hence the name, but we were lucky to enjoy a feast of Southeast Asian food at our smoko. It was far more exciting than the casse croûte of baguettes and coffee in the Loire Valley.

The Land Rover had just pulled away from the farm when the confusion began. I heard Sev, a few rows away, mutter, "Um, what the fuck?" Raphael had stopped picking. Sam was standing over a bin, which was already half full. He was holding up a cluster—and it was not a gorgeous, black-skinned Gamay bunch. Instead, it had barely gone through *veraison*—meaning, it was partly still green.

Much of the Gamay was partly green, apparently. Everyone was perplexed as to why Wildman had asked us to pick here, on this day. He'd measured yesterday, so surely he knew that ripeness and potential alcohol were very low. Or had he only taken his sample from certain parts of the row? What would happen with all these grapes? Could we make drinkable wine? We continued picking, but Sam came up to me with his phone to his ear. He was calling Wildman to let him know what was happening.

"I think he might listen better to you," Sam told me.

I took the phone and held it to my ear with a sticky hand.

By then, Wildman had made it back to the farm. I could hear the pressure washer running loudly in the background, cleaning the press.

"Hey, sweetie, um," I looked around at the pickers who were eyeing the green bunches warily. "Why are we picking the Gamay today? It's not ripe. Everybody is really confused."

Wildman was brief in his reply. "Yep, it's fine. Just keep going."

Was he too distracted to care? I hung up and gave the phone back to Sam, reporting the order.

It took another phone call to Wildman before Sam was satisfied—finally, Wildman said to just pick the darker clusters and leave the other ones for another day. We slowly went from one vine to the next, hunting out the ripe grapes. I'd heard of early harvesting for lean wine, but this seemed extreme. Raph, Sev, and Alex seemed annoyed, and the pickers did not love the slow pace. They were paid by weight, so today would not be favorable to their income.

When the sun had risen high in the sky, we stopped for smoko. On a grassy patch near Rob's metal workshop, Sam and his crew laid out a colorful straw mat. The menu changed every day, and we eagerly watched as the food came out of coolers from the back of Sam's van. Today the crew had prepared rice noodles flecked with all sorts of green herbs grown in their backyards, as well as fried chicken wings and sticky rice, tiny raw eggplants and hot chiles, lemongrass sausages and pink fatty pork sausages. The crew watched with smiling eyes as Raph, Alex, Sev, and I grabbed handfuls of the warm sticky rice, formed little balls, and sunk them into the saucy dishes, eating until we had to lie down under a tree, sweating. We might regret having overeaten, especially as we returned to the vineyards for many more hours of picking, but it was not enough to stop our wide-eyed gluttony.

If there's one grape that's at the heart of the natural wine movement, it would have to be Gamay.

In the Southern French region called Beaujolais, Gamay is the exclusive red variety permitted for winegrowing. For the French wine elite, the Beaujolais region is considered to be far less prestigious than Bordeaux, with its age-worthy blends, or Burgundy, where Pinot Noir claims extremely high prices. There's a history to this disdain for Gamay, which is an extremely versatile and pleasant variety.

Several hundred years ago, Gamay was planted throughout Burgundy's slopes. But the Duke of Burgundy considered its acidic flavors so vile that he evicted it from that region in 1395, declaring only Pinot Noir fit for growing. The French may have progressed beyond aristocratic values with their eighteenth-century revolution, but they kept their snobbery toward Gamay mostly intact, save for the annual

Beaujolais Nouveau celebration in November, when the wine is released just after fermentation has completed and drunk by mobs in the streets of Paris.

In the late 1980s, the Beaujolais region had a particularly bad reputation. Winemakers were known for producing nothing but simple, boring, early-release wines. They often used a method called "chaptalization," which involves dumping processed sugar into fermenting wine to make it more alcoholic, to improve its commercial prospects. In this climate, a man named Marcel Lapierre began experimenting. He wanted to see whether the wines could be improved by eliminating preservatives—namely, sulfur dioxide, or SO_2 (colloquially, "sulfur" or sometimes "sulfites," or in British English, "sulphites").

Lapierre worked alongside local chemist and winemaker Jules Chauvet, who today is considered a forefather of natural winemaking. Chauvet had tried some initial experiments in his makeshift winery. When Lapierre's sulfur-free experiments succeeded in producing tasty and unflawed wine, he spread the word to some friends that wine made from grapes, and nothing but grapes, was possible. Of course, they knew that their grandparents had made wine this way—but it had been drilled into them that preservatives (and other additives or manipulations) were vital in making commercially viable wine. Without preservatives, the belief was, wine would become vinegar. Chauvet and Lapierre proved that technical prowess and extreme care could make for a perfectly fine, drinkable, vin nature—natural wine.

Over time, Lapierre and three other Beaujolais vignerons, including Jean Foillard, Guy Breton, and Jean-Paul Thévenet, came to be known around France and later the world for instigating a return to quality winemaking. Their US importer, Kermit Lynch, dubbed them the "Gang of Four," and the name stuck, although one or two other winemakers were also involved in these early experiments.

Creating this change was a long, uphill battle for the Gang of Four. As they reduced their usage of chemical treatments in the vineyard, people called them crazy for exposing their vineyards to the elements. The notion of fermenting and bottling wines without sulfur seemed risky and foolish. But they persisted, and over time, they found their audience. First, it was in Paris—people like the legendary chef-sommelier team Raquel Carena and Philippe Pinoteau. Their bistro, Le Baratin, in Belleville, was one of the first restaurants to support makers of organic, preservative-free wines. Le Baratin paved the way for other restaurants to do this through its overall no-frills, ingredient-focused approach to all things on the table, inspiring the current generation of young Parisian chefs, and it has stayed the course over time as a landmark for dining and drinking well.

Over the years, vignerons around France heard of Marcel Lapierre and traveled to meet and learn from him—René Mosse had mentioned to me how "meeting Marcel" had been an inspiration. In this way, his influence fanned out over France as more and more producers began trying to make natural wine. Thanks to the Gang of Four and others around and after them, the Beaujolais region has come to be valued for its terroir almost as much as Burgundy, and now wines made in the ten *crus* (growths) of Beaujolais are considered age worthy and claim high prices.

Gamay is a variety that quickly entices with its versatility; it's pretty and light on its feet, but there can also be depth and structure to a Gamay wine. In Rob's Hahndorf vineyard, there were five different clones of Gamay grafted onto the vine rootstocks, a "clone" being essentially a copy of any given parent variety, raised from cuttings taken from a "mother vine" in a nursery. Each clone varies in skin and juice color and flavor. But these clonal plantings were interspersed, so it was impossible to know which was which. It amazed me that, just by tacking on cuttings of Gamay, a vine that was previously Sauvignon Blanc completely changed its identity.

And thanks to Wildman's grafting work, we were now making some of the very first Australian Gamay.

But how would it turn out, having been picked so green? I wanted to trust Wildman, but I doubted his judgment here.

When he returned to the Hahndorf vineyard to pick up the grapes and collect us interns (although I didn't quite think of myself as an intern, per se), Wildman inspected our Gamay harvest. He seemed pleased. We, on the other hand, were exhausted and even more fatigued knowing that there would be processing to do back at the winery, and then the next day we'd repeat the whole thing. It was halfway through vintage, and our bodies were feeling the strain.

Back at the winery that evening, Sev and Raph and I did punchdowns on the Pinot Gris, using our body weight to soak the cap that was forming on top of the fermenting grapes. If it became dry, it could upset the yeast and bacteria in the wine, and all sorts of flaws could occur.

Meanwhile, Alex forklifted the Gamay into the air and dumped it into a large stainless-steel tank, where it remained for several days, gaining color and tannin and flavor from its own skins.

Dinner was tired but also cheery. "So, do you have a plan for that Gamay? There's quite a bit of green in there," I asked Wildman as Raph distributed plates of spaghetti, which Sev and I had cooked. We tried to be egalitarian about meal prep, everyone helped if they could.

Wildman leaned back. His steaming plate was yet untouched, although I hadn't seen him eat all day. "I want a nice, low-alcohol wine," he said. "Maybe it will work out. Every year, it's different, and I never know what I'm doing." He shrugged. That was winemaking—a completely different crap shoot, year after year.

When we returned to Rob's the next morning, Wildman said to go ahead and pick all the Gamay, even the unripe ones.

"He's gone totally nuts," I muttered to Sam as we bent over to grab buckets full of grapes from below the pickers. They worked so quickly, we could hardly keep up with the forty-pound buckets. "How is the wine going to be? It can't be good, if the grapes aren't ripe, can it?"

Sam shrugged and managed a diplomatic expression. He was in a difficult position sometimes, trying to keep the pickers happy while following Wildman's instructions at the same time.

For two days, we picked other sections of the vineyard, and then we returned to the Gamay, which to my relief had ripened fully over the past few sunny days. Some of this fruit would be mine for the making. I hadn't really thought about what I wanted to do. Munching on clusters, including the stems, as Wildman had instructed me, I concentrated on the flavors. The stems were definitely "green," meaning unripe and somewhat vegetal tasting, but I didn't like the idea of taking them away from the grapes. When I ate good artisanal cheese, I always enjoyed the rind along with the creamy part, appreciating how they flattered each other. So, I would treat grapes the same way. Destemming grapes was a popular approach, I knew—possibly it ensured a safer fermentation, with less oxygen. But I wanted to let the grapes be whole. It wasn't a "winemaking" decision so much as an intuitive one.

That evening, as the car rolled onto the farm, with Alex at the wheel, Wildman ran out of the winery and knocked on the car window. "Come in, come in!" He motioned for us to follow. We all really needed a shower, but we joined Wildman in the shed, at the stainless-steel tank where our first Gamay pick, the semi-green one, was starting carbonic fermentation. He held a Zalto glass with the stem broken off under the spout of the tank and opened the valve, releasing a bit of light red juice. Then he handed the glass to Alex, who took a sniff and a taste and passed it around. When it got to me, I noticed that my fingers were almost entirely black from picking.

The juice, which wasn't yet wine but was showing some of its future wine-self, was actually good. It did have green flavors, but also some black pepper, curry, raspberry jam, and crunchy red apple. It was nice.

Really nice. Wildman was beaming with pride. "And you didn't even want to pick this!"

Did he know ahead of time that it would work out? Not exactly. But he had a hunch that the wine would have just enough sugar to ferment properly. He anticipated that it would turn out to be a fantastic light spring drink. It was just a matter of responding to the moment. And maybe, I suspected, a bit of ego factored in. Wildman was a strong-minded person. Even if he didn't know with certainty what he was doing, and couldn't back up his ideas academically, once he got a notion in his head, there was no point in pushing back.

I was making my very own Gamay.

Only six months earlier, I'd felt like a buffoon in the Mosse family winery. So what business did I have making wine now? None, exactly, except that for whatever reason, Wildman really wanted me to do it. And he seemed to believe in me.

By the time we were dealing with the green Gamay, I'd already started making one wine: it was envisioned as a pét-nat of Chardonnay and Pinot Noir, a "baby Champagne," based on two varieties common in Champagne. The grapes had been pressed already—the Pinot had been on its skins for just forty-eight hours, for a little color, in a large wooden barrel with the lid cut off. I'd shoveled those grapes into Wildman's press myself. The Pinot and Chardonnay had been happily fermenting in two separate barrels for a week now. Ideally we'd bottle them before they fermented dry, to make sparkling wine.

With this Gamay, I had a choice: I could opt for a "carbonic" fermentation, as Wildman had done. For this technique, common in Beaujolais, we'd dump the grapes into a tank and seal it, so that fermentation would originate from within the grapes. Or I could do a "traditional" fermentation, meaning the grapes would rest in a large open-top wooden barrel for a few days. In this method, I would jump on them to get the juice going—in my bare feet, as winemakers have done for thousands of years—and do twice-daily punchdowns to ensure that the cap was wet. I opted for the latter—it seemed less risky, since I'd be able to check on the juice every day to see how it tasted and therefore make an educated decision about when to press. It also seemed more active and fun than just waiting while the grapes sat in a tank.

With the forklift, Wildman dumped two bins of Gamay grapes into a large open-top wooden barrel, the same one I'd used for the Pinot Noir. I stripped off my pants and climbed in. Ruby-shaded liquid appeared around my ankles with each cluster breaking open as I jumped up and down.

Jump, jump, jump—it was exhilarating. Across the winery, Sev and Alex were prepping some barrels for use. They lowered them onto a rotating device, then turned on the pressure washer to run hot water through the barrel. The pressure washer was extremely loud, though not as loud as the Metallica they were blasting from speakers. Amid all that, there was the noise of the press whirring. And everywhere in the shed was Wildman's chaos: a jackhammer resting here, barrel fittings piled up there, a ream of his fancy label paper carelessly left in a bucket that was half full of various assorted rubbish.

I couldn't hear myself think. Down at my ankles, there were a few good inches of fresh juice covering the grapes—enough. I got out of the barrel, wiped off my legs, and redressed. Over the top of the wooden fermenter, I draped a clean bedsheet, then pulled an elastic band around the sheet to create a seal.

Then I rinsed my hands off with the hose. I needed a break from the cacophony of this shed. Noting that everyone was preoccupied, I decided to take a moment to myself. I grabbed a bucket and found a pocketknife and headed down the driveway toward the veggie patch.

Alfie and Lulu appeared at my heels. I walked carefully on the driveway, which was steep and paved only with gravel, so as to not slip. Once at the patch, I headed into a long row of kale and began to slice off bundles, thinking forward to lunch. The dogs tumbled around beside me. It was peaceful in the patch, with birdsong and a light breeze.

Working in the winery all the time, frankly, didn't appeal to me—it was too noisy, too cluttered with machines, too frantic. I didn't want to be Wildman's protégée, anyway—I wanted to continue publishing *Terre* and writing about wine. Just before vintage, I had turned in a feature story for *Imbibe* magazine about winemakers collaborating on special bottlings. With new directions in freelancing and publishing *Terre*, my career seemed to finally have momentum.

With a bucket full of kale and some early tomatoes, I marched back up the hill. As I was passing by the little green shed, which Wildman had used to store and label wines before constructing the newer, larger shed, a Subaru Forester pulled up and parked just above the Pinot vineyard. It was Oli and Bree, the chefs from the Aristologist. They were hoping to join us in the vineyards in the next few days to make their own wine, just for home consumption.

"Hey, how you doing?" They greeted me, and, as I answered I'd just been jumping on some Gamay, they popped the trunk of the car. Piece by piece, they pulled out a set of wooden doors and a metal base. It was a simple, vertical, manually operated basket press, the kind that nearly all winemakers would have used before the invention of electric presses.

I followed them into the shed, and they showed me how the basket press worked. Looking out over the Pinot vines and the veggie patch down below, with the forest where kangaroos roamed wild just beyond, and the noise from Wildman's shed only distantly audible, I could suddenly hear myself think. And I was having a very clear vision of myself, in that shed. An idea came to me.

Just four days later, I tasted my Gamay and judged it ready to press.

"I think it's had enough time on the skins to develop flavor and tannin," I told Wildman confidently. I explained my plan: I would be pressing the Gamay with nothing but the humble basket press the chefs had brought over, using my two hands and a few buckets to get the job done.

Wildman loved the idea. "I'll bring the fermenter down," he said, hopping on his tractor. He instructed me to carry down a large, vertical sieve and a long red plastic hose. I also grabbed a shovel. Then, with careful movement of the forks attached to the front of the old green tractor, which terrified me with its enormous wheels, he lifted a half ton of grapes.

Down in the little green shed, I pulled the sheet off the large wooden fermenter and, standing on my tiptoes, peered in over the edge. The fermenting grapes inside were just covered by a layer of free-run juice, mostly created by me jumping on them. Aromas swirled around—funky yeast, cherry cola sweetness, pungent carbon dioxide.

Using the tractor forks, Wildman raised the fermenter. He hopped off the tractor and dropped the large sieve into the grapes, then the hose inside that sieve. He began sucking at the other end of the hose, a job I'd tried and found nearly impossible—it required immense lung power. This was the best way to siphon off the free-run juice, without using an electric pump.

"There you go. Anything else you need?"

It was sincere, but I could also see in Wildman's face that he was thinking about a million other things that needed to be done for his wines. I thanked him and sent him back up the hill. And then I was alone—except for Alfie and Lulu, who lay just outside the shed, looking for kangaroos to chase.

With a shovel and a bucket, I began transferring the semi-fermented grapes to the basket press. Their powerful scent—yeasty, fruity, sour, earthy, all at once—energized me. It was hard, sweaty work, and my shoulders soon ached. But then I saw the first rivulets of ruby-tinted juice streaming out of the press and into a bucket that sat on the ground, and I felt a rush of new energy. Listening to the sound of my own breath, I filled the press. Once all the grapes were inside the basket, I lifted the heavy wooden semicircles, one by one.

I maneuvered the semicircles to form a closed layer on top of the press. Then I had to bring down the actual press by raising and lowering a lever so it would lower onto those semicircles, which clamped down on the grapes.

Click-clack, went the lever with each pull.

Juice flowed faster and faster into the bucket, and soon it was full. I switched it out for an empty bucket and poured the fresh juice into a barrel that was resting on its side, on a pallet. The juice smelled heavenly and its color was ethereal.

I'd begun making the first wine I felt I could truly call my own.

– Eleven –

Persephone's Fall

Working the basket press was no joke. I had to use all of my upper body strength to pull the lever down, putting weight on the grapes. And getting the last bit of grapes and juice at the end of the fermenter was an unexpected challenge—the vessel was really tall, and there was no way for me to reach its bottom. I could get inside, but there was nobody for me to hand the bucket to. Alone, this was really hard. I could go up to Wildman's shed and ask for help. But my stubbornness prevailed.

I hugged the barrel and used all my strength to tip it onto its side. Then, I cautiously removed one hand and, holding it steady with the other, scooped up the last grapes with the bucket. They went into the press. Then, fueled by adrenaline and a cold beer, I spent a good hour working the lever, slowly building up pressure on the grapes, capturing the juice running out from below, and transferring it bucket by bucket, into the barrel.

It was probably the silliest way anyone has ever unloaded a fermenter. But I did it. The barrel was nearly full of fresh Gamay juice. My muscles were on fire. And now I had to take the press apart, remove the hefty "cake" that had formed from all the mashed-up grapes, and hose down the doors and basket of the press.

At dusk, I clambered wearily up to the house and awaited my turn to shower. Raph, Sev, Alex, and Wildman had also had a long day. Wildman was in the kitchen, grating carrots for our dinner salad. I wrapped my arms around his waist and exhaled into his shoulder blades.

"Have fun down there?"

"Yeah," I said. "You could call it fun, I guess. How was your day?"

"We got the press running," he said, looking through the window toward the shed. The press sat on the concrete platform, whirring loudly. "Should be finished by midnight. Don't think we'll do another press load tonight, we'll see."

Then he mentioned that we'd be picking Cabernet Franc the next day. Did I want some?

I paused. The plan had been to make one barrel of wine. But I felt I was just getting the hang of fermenting and pressing. Why stop now? It would be nice to have more than one varietal wine, I thought, to see the difference in how they turned out and also to have more to share with the world.

By the time I finally got into the shower—which at that point, of course, was cold but better than nothing—I'd agreed to also make Cabernet Franc.

Picking the Cab Franc was less dramatic than picking the Gamay—it was clearly ripe. Fresh off the vine, the fruit tasted strongly of pyrazines, a flavor note that resembles green bell peppers and a hallmark of this variety, native to the Bordeaux region.

After picking all morning and afternoon, I stood in the green shed before two bins of Cabernet Franc—a half ton in total. Wildman had brought it down with the tractor while I washed the day's picking buckets, scrubbing grape skins off with a sponge.

I shoveled the grapes into the same wooden fermenter I'd used for the Gamay. Then I climbed in. For twenty minutes, I jumped on the grapes. While jumping, I absentmindedly picked through the clusters, removing stems as the berries fell off them, examining the quality of the grapes.

As I jumped on the grapes I thought back on the past year—much had happened. I'd suffered a few romantic disappointments, packed up my things, and left the city that had been my home, my comfort zone, for nearly eight years. I'd said goodbye to friends. Had traveled to Georgia and met Wildman. I'd shown up in Paris and gone through that terrible heat wave, alone, in that awful Airbnb, with only Wildman's delivered flowers to brighten my day. Worked the harvest in the Loire. Bonded with Gaba through our strange Paris lives. Launched a print magazine.

I climbed out of the vat of grapes and examined my work: they were nicely broken and juicy now, ready to ferment. I was sweating and my muscles ached. But within, I felt cleansed and calm. The tumultuous past year and my attraction to Wildman were beginning to reconcile, although I wasn't sure where it was all leading.

But being spiritually lost on the farm seemed fine in a way that being distraught in New York or Paris had not. Here, we ate healthfully, with vegetables grown a stone's throw from the house, and drank plenty of natural wine. All around us, there was untouched nature. In the wild eucalyptus trees and gigantic pines, all sorts of birds spent their days. Kangaroos pranced around on the hillsides and made paths into the woods. The sheep roamed on the hillsides near the house, although the dogs barked at them so much that they had been breaking out of their paddock and wandering into neighbors' vineyards, which meant Wildman was receiving angry calls. There was a chaotic sort of harmony among all these living things. I was slowly coming to see why, in a world beset by ecological crisis and overcrowded cities, living rurally was an attractive option.

Four days later, after a few more sessions of jumping on the grapes and doing punchdowns every morning and night with my hands, I tasted the Cab Franc. It was ready to press. I walked up the hill to tell Wildman and grab the equipment I needed.

Wildman's shed was buzzing with activity. Alex was forklifting grapes into the press, while Sev was sweeping up grape clusters below. A few other interns were there that day, and they were washing out picking bins. I saw that Raph was struggling to siphon juice from one vessel to another and empathized with his plight. Wildman was running around amid all of it, supervising and tasting and smelling. I told him I was ready to press the Cab Franc.

"I'll come down," Wildman said. While he got the tractor going, I dashed around, hunting for the funnel, a clean barrel bung, and a few other items. He loaded an empty barrel and the sieve onto the tractor forks, I jumped onto the side of the vehicle, and we rode down. Wildman smiled at the sight of me, hanging onto the tractor door. He seemed really happy—he was in his prime, making split-second decisions, working with his hands, mentoring all of us in the art of natural winemaking.

As Wildman unloaded the sieve and barrel, I carefully lifted the sheet off the fermenter to reveal the mess of semi-fermented grapes inside. First, there was the free-run juice to deal with. Wildman began hoisting up the fermenter, with the tractor forks—gravity was our friend, since we didn't use pumps to move liquid.

"Want to try the siphon?" he asked.

"Sure, I'll give it a go." I began sucking on the hose with all my might. Wildman leaned against the wall and I heard some hip-hop music coming from his phone. It was a South African group he liked—something he'd listened to as a kid back in Johannesburg.

This was now my third or fourth time trying to siphon wine, and I continued to find it impossible. My chest hurt. But no matter how much I tried to inhale deeply, to get that juice flowing, it wouldn't come. I was exasperated.

"I can't fucking do it!"

Wildman looked up from his phone. "Want me to?"

I shrugged and stepped aside, annoyed. It felt like a replay of my attempts to be "helpful" at the Domaine Mosse winery. Within two minutes, Wildman had established the siphon. He handed me the hose and I "walked it out," lifting it above my head and twisting this way and that, moving the juice through until it reached the barrel.

"Thanks for your help," I said, out of breath.

"No worries," Wildman replied lightly.

The barrel was filling with the free-run juice. Now I could begin shoveling the grapes into the press. Wildman also grabbed a shovel, and we worked side by side.

Pressing my Gamay, I'd felt a sense of flow. It was even meditative. But now I was distracted. The music, for one thing, was really bothering me. It was like some mediocre hip-hop that a bad radio station might have played in the '90s. And Wildman's energy felt really off. He was shoveling quickly, clearly in a rush to get through the task. Up at his shed, there was plenty going on. But his hurrying was putting me on edge.

"You know what, I'm actually fine," I said. "I got this."

"It's all right," said Wildman, heaving a shovel full of grapes into the press.

But it wasn't all right. The whole point of the basket press was that it didn't require any machinery and could be operated by one person. It was slow, meticulous, by definition. This process was the reason I'd agreed to make more wine, beyond just the Gamay—it was something I enjoyed. It felt like an intimate, intuitive approach to making wine. The means were just as important as the end.

"No, really, I'm OK—please. I mean, I'm sure you have other things to do."

He stopped, glaring at me. "Sorry, I was just trying to help."

I snapped. "Well, I didn't ask you to help, did I? And what is this music? Can I be allowed to choose my own music when I'm making *my* wine?"

"If you don't want my help then I'll go away!" Wildman was fuming.

"Actually, that would be great. Thank you."

In a few minutes, he'd backed up the tractor and was headed up the hill. In the silence of the afternoon, with Alfie and Lulu keeping me company, I slowly worked the press. When about half the grapes were in, I had to take off my boots and climb up into the press to squash it all down, to make space for the rest of the grapes. Then I loaded on the semicircles and heaved the lever up and down.

After I unpacked the press to reveal the cake the dried grapes had formed. I cracked it open with a plastic pitchfork and scooped it all into an empty bin, then hosed everything down. The whole thing took about four hours.

It was the most exquisite exhaustion. And it was all mine—or mostly.

I had grown up in a household with a single mother. The notion of a woman's independence was deeply ingrained in me. I needed my own space; it was simple as that.

Wildman had offered me some of the Sangiovese we were picking the next day. I resolved that, when it came time to press that Sangiovese, I would do the entire thing without *any* assistance.

The overall energy on the farm transformed with the progression into late autumn. We'd done around 80 percent of the picking, and now there were other matters to focus on. We had to constantly monitor the ferments to be sure they were safe and healthy. This meant ensuring that there was no VA (volatile acidity) or the similar-but-different bacterial problem known as EA (ethyl acetate). To detect and prevent these, we scanned for traces with our noses during punchdowns. We also had to make sure barrels were full and top them up with extra wine if need be. Wildman was adamant about regularly tasting anything that was still on its skins, trying to find the exact right moment for pressing. Too early meant not enough flavor. Too late meant overextracted, inelegant wine.

As the various ongoing ferments began to feel somewhat under control, Wildman began to turn his attention to the farm at large. On a day when strong, cool winds pierced the warm morning sunlight, he emerged from the depths of the winery wearing what looked like an astronaut suit. He wasn't going to the moon but rather to the beehives, which claimed a little corner of the forest, just beside the veggie patch.

Wildman pointed a smoke blower into the hives, agitating the bees so they'd leave their home. He then reached in and removed the wooden structures where the bees had built their honeycombs. I watched from a safe distance. Years earlier, I'd been stung a few times while rock climbing and had gone into anaphylaxis. During this vintage, I'd already been stung and nothing major had happened, but I wasn't taking any chances.

But once the honeycombs were in the winery, I offered my help in spinning the honey. For hours, I sat on the floor, rotating the combs so that the honey dripped off through a sieve into jars, mesmerized by the simplicity of the process as well as the complexity of what these bees had done.

Later that week, Wildman announced that we should all grab buckets and follow him along a path that led from the house to the neighbors' vineyard. We walked past their vines, down a steep hill until we came to a grove of olive trees.

"Koroneiki and frantoio there," Wildman said, pointing to trees with, respectively, medium-sized and very small green olives. "And those are my prized kalamatas." He gestured to a tree with tear-drop-shaped fruit. Instinctively, Alex picked one off the tree and popped it in his mouth—just as we'd do in a vineyard. He grimaced at the repulsive, sour flavor.

Wildman laughed. "Not so pleasant until we've brined them."

Through the leaves of the olive tree, sunlight shimmered and danced on our forearms as we navigated the branches to pluck olives one by one, dropping them into the buckets—we soon filled five. Wildman seemed really energized. He'd cut his own hair and given himself a nice shave, and I loved being able to see the contours of his cheeks and neck. He was incredibly handsome when he smiled or laughed and his eyes were framed by the crinkles of time.

"I planted these trees five years ago," he said. "But it's the first time I've harvested them."

"Why? Were they diseased?" I replied.

He answered, "Never had the time before. Always too much work to do."

Wildman's journey to this present moment—building that new winery shed, making wine with minimal assistance each year, hacking through an acre of stubbornly invasive blackberries to plant veggies for his restaurant—was something I'd not much considered. Wildman had never had it easy, I thought as we lugged the buckets of olives back to the house, where we would put them in water to start the fermentation process. He had put in many years of hard work to have this life.

And here I was just stepping into it, making wine, mothering these puppies. I didn't even know if it was the life I wanted. Maybe I was leading him on. Maybe I was taking advantage of his love and generosity, when all I really wanted was to be a grouchy, broke writer in Paris, where I would drink at every wine bar, searching for a love that didn't exist, denying my own loneliness.

I stood alone in my shed with my familiar friend, the wooden fermenter. This time, it was full of Sangiovese grapes. The chefs' press was assembled and ready to go. Now, I had to confront the task of dealing with all the free-run juice in this Sangiovese, unassisted, as I'd planned.

There was a good deal of liquid sitting in that fermenter. The native Tuscan grape Sangiovese, it turned out, was a particularly juicy variety. Wildman and I had laughed, as we harvested these grapes from Rob's vineyard in Hahndorf, that they resembled Krusty the Clown's unruly hairstyle—large bunches of grapes with tufts on both sides. For this Sangiovese, I didn't even have a previous vintage to taste for reference, as Wildman had grafted it onto the vines only a few years earlier, and it hadn't produced enough grapes until now.

I would have to carefully transfer the juice along with the berries, bucket by bucket, to the press. I put on Sylvan Esso at full volume, gulped some water from the hose, and got to work. It went perfectly fine, for a while. Within an hour, I'd loaded the press with nearly all the grapes. But there was still some liquid remaining at the bottom of the fermenter. I managed to tilt it on its side, resting it briefly on the picking bin, and peered inside. It looked like at least thirty liters. No way could I leave that juice and those grapes in there. But the barrel was very tall, and I couldn't reach the bottom.

I crouched and brought the heavy barrel toward the ground. The juice crept toward the opening. With my feet, I inched a bucket closer, and then let the fermenter go nearly all the way down. But I missed my target entirely. Very quickly, the juice rushed out past the useless bucket, and my boots and the floor of the winery were covered with Sangiovese.

"Shit! Shit! No!" I sat on the floor of the shed, my head resting in my hands. I was exhausted from the early mornings and late nights. Everything hurt. I'd wasted some of the precious Sangiovese that Wildman and Rob had worked so hard to grow, that Wildman had been so generous to offer me.

All because of my damn pride. Because I'd wanted to do it without help.

Escargots, baguettes, and cheap beer were on the screen in front of my eyes.

"God, I wish you were here right now to share this with me!" Gaba said.

I exhaled my cigarette toward the valley below the house and looked at her. Gaba was wearing sunglasses outside a café. Her recounting of her daily life in Paris, in our regular video calls, was starting to make me feel nostalgic for Europe. I could almost feel the wicker seat and longed to be there, casually watching people around us, observing their sense of style, the way everyone looked so good and projected such confidence.

I'd been wearing the same jeans and sweater for three days, and my new Rossi boots were now well worn in.

Gaba was finally past the first trimester—the morning sickness had been rough, and she was elated to be done with it. It was early spring in Paris, and she and Dan had been settling into their new apartment in a very fashionable district.

"I saw a place the other day that would be perfect for our bar!" she continued.

It wasn't that long ago, Gaba and I had sat in my bathroom in Brooklyn, smoking and plotting our Parisian lives. Was I cruel to not stay true to our original vision? Or was I stupid to not see the fulfilling life that was unfolding right before my eyes in Australia?

"Tell me more." I listened as Gaba described a failing restaurant, run by an elderly man, not far from the market we'd frequented in the 12th arrondissement. It was a bit dark inside, she said, but had potential. Nice wooden floors. A workable kitchen.

"When will you be *back* here?" She wanted to know.

Wildman and I had been making plans to travel together to Europe for a wine event in Copenhagen and then visit some producers in France and Italy. But it was a round-trip flight.

"We'll be in Paris for a week in May," I said, "but then I have to come back here to bottle the wine and all that. I'll be back in September for the baby's birth!"

"Oh, so in September you'll move back here?" Gaba was optimistic and persistent. "Perfect—here we have subsidized childcare starting when the baby is six weeks old, and by then maybe we'll have an investor on board, and we should be ready to start renovations."

Renovations. We had hardly made any kind of a business plan. Nor was there an investor on our horizon. But anything was possible, I supposed. I'd started a magazine with two friends and a Kickstarter. Was a wine bar so different?

I blew kisses into the screen and said I had to go back down to the winery.

It hurt to say goodbye to Gaba. I didn't know if our friendship would ever be as it was, those weeks we'd lived together in Paris. Also, I had yet to find anyone in the Adelaide Hills to whom I felt close.

There were plenty of local women who intrigued me. There was Monique Millton, of Manon Farm, the natural winery up the road from us. The other night, we'd driven up there, high upon a ridge above the hills and valleys. They had a nicely sized vineyard they tended themselves, as well as their own biodynamic preparations and compost, and olive trees, and a veggie garden—it was impressive. Monique's partner, Tim, who had DJed at Wildman's shed party back in December, was a former chef, and she hailed from a family of biodynamic winegrowers in New Zealand. We'd enjoyed the sunset while sipping on their Savagnin wine, a rare variety in Australia that was originally believed to be Albariño when planted, and nibbling on Monique's homemade sourdough and fresh ricotta.

Monique seemed really busy with their two-year-old son and the ongoing work at their farm. To me, she seemed like a real farmer, while I still felt like a fake.

There was also Lily, a spunky woman who worked at the organic restaurant in a nearby town and who was always whipping up herbalist concoctions like calendula salve. There was Sarah, whom I'd met that night at 10 William Street, after Rootstock—she worked long hours at a winery restaurant in the Hills but often appeared at the Aristologist post-shift for a nightcap. There were, in fact, so many women who led interesting, busy lives in the Hills. But the Hills weren't like Paris or New York—social worlds were very slow to expand here. The result was, after three months of being in Australia, I didn't feel close to anyone there, except Wildman. Never in my life had my partner been my sole confidant, the only person with whom I could enjoy spending time.

Sev and I had started out vintage as friends, but in recent weeks she'd given me the cold shoulder. I tried to speak to her about it, but she was dismissive. I suspected the abrupt change in her attitude was related to my "status" as Wildman's girlfriend.

I missed Gaba, and I longed for the easygoing life of a vibrant international city, where people came and went and conversations flowed.

I'd known back when I visited my mother in Virginia how I'd name my nascent wine brand, although at the time I believed it would be only one barrel. The book of Greek myths, and Persephone's story of tumbling into the world "down under," had inspired me. Of course, her version was much darker, as she went unwillingly. I appropriated her name for my wine brand in a somewhat less literal way.

This tension between sunshine and darkness, between hemispheres, between wanting companionship and a desire to be free, and the narration of the agricultural seasons, seemed like a perfect analogy for me making wine in South Australia with Wildman after trying to move to Paris.

– Twelve –

Natural Selection Theory

"Finally—an afternoon off!" Sev muttered quietly yet audibly to Alex and Raph. The three of them were crammed into the back seat of the Land Rover, smoking out the open windows. After weeks of picking and processing grapes, everyone was exhausted. For a little break, Wildman called up his old friend and protégé, James Erskine, and we had an appointment to taste his latest Jauma wines in his shed, just down the road in Basket Range.

We drove along a very windy, narrow road, passing the house where Aaron and the Aristologist chefs lived, swerving around the bend at the local town hall, where an Italian woman taught weekly yoga classes. At the town hall, there was also a tiny, adjacent post office—open for just two hours each morning—and fire house, where devoted volunteers worked to monitor the often-dangerous bushfire situation through the warm months. It occurred to me as we drove that, since arriving to Basket Range, I had not seen a single flat stretch of land.

Jauma was a longtime favorite of mine, which I'd enjoyed in New York as well as Paris. I appreciated the fresh, acidity-driven quality of his wines, including several Grenache and Syrah red wines, as well as a lightly fizzy pét-nat Chenin Blanc that had livened up several afternoon picnics. Jauma wines were visually distinct—they came in a clear, weighty Champagne bottle, rather than the traditional brown glass and curvy "Burgundian" shape preferred by most winemakers. They had simple crown caps rather than corks. Their labels featured illustrations by James's two young children, with very little information about the wines—usually the variety, the name of the grower, the year, and nothing else.

"Oh hello, there." James leaned his head in the window of the Land Rover on Wildman's side. We all perked up and smiled back—James seemed to radiate kindness and positive energy, and it was infectious. His dark, gray-streaked hair was pulled back into a bun, and there were handsome crinkles around his eyes, which deepened as he gave us all a welcoming smile and landed a big smooch on Wildman's cheek. We climbed out of the car like a clan of sunburned, bone-tired clowns, ready to drink and learn and enjoy the fact of not working.

"All right, let's try some things," James said, wielding a handful of wine glasses. We followed him around the shed as he knelt below each tank full of freshly pressed juice, poured out tastes, and spoke about the vintage. Although the vineyards he sourced from were not distant from ours, the microclimate was different. Everything ripened earlier in the McLaren Vale. That's why it had varieties like Grenache and Shiraz, whereas the Burgundian varieties that preferred cool climates were planted elsewhere in the Hills. The juice we tried was not yet wine; it was nearly finished with fermentation but still had a touch of sweetness left. We had to envision what it would eventually be like.

James disappeared behind a door. He came back a few minutes later holding several bottles, which he lined up atop a barrel, moving, as he always did, in a sprightly way, like a forest fairy.

"Well, that's a trip down memory lane!" Wildman was eyeing the bottles.

"Yep, here's one of our earliest, I reckon." James pointed to a strange-looking, brown bottle, with a clasp that could be flipped off. It wasn't a wine bottle, at all—you'd use it to serve water, more likely. The label said simply, "Live Red, 2011."

"Why don't we start with this one," said James. He and Wildman exchanged quiet talk between them about how vintage was going, while uncorking a 2012 wine labeled "Mère Syrah." I noted that it used the French name for the grape, rather than the typical Aussie version, Shiraz. The wine had a silky richness to it and definitely seemed like it was meant to impress.

"It's strange to try a wine from you that had sulfites added to it," I remarked to Wildman. It was obvious to me without even asking that sulfites were added to the wine—I couldn't describe exactly why I knew, but my palate could tell.

He shrugged. "Back in 2012, we were adding around 40 parts," he replied, meaning 40 parts per million. "That was pretty radical, in Australia, for the time."

I nodded—I knew that most mass-produced wines found in grocery stores often had 90 ppm or more. Many fine wines were dosed with between 40 and 70 ppm of sulfites. The wine fair, RAW WINE, which promoted itself as a natural wine fair, allowed up to 70 ppm of added sulfites, as long as other criteria regarding viticulture and vinification were met.

In a short while, we opened the 2011 "Live Red," a red blend, the first no-sulfites-added wine that James and Wildman ever made.

We silently inhaled this seven-year-old natural wine's aroma. Dark cherries laced with cloves and cardamom greeted my senses. It was showing excellently, despite all the time that had passed. I'd tasted a few nearly decade-old sulfite-free wines before and marveled at their vibrancy and perfection, but it never ceased to amaze me that preservative-free wine could age so beautifully. We were about to have a tasting designed to take us into a liquid history of the natural wine movement in Australia—and all the goofiness, camaraderie, and tragedy that marked its beginnings, for Wildman, James, Tom Shobbrook, and the late Sam Hughes.

It was January 2010, and the three South Australians were traveling across the windy, jagged coastline on the Great Ocean Road, taking the long way to Melbourne. To break up the thirteen-hour trip, they stopped at James's family farm, where Tom and Wildman pulled their rifles and a speargun out of the back of Wildman's brand-new Land Rover. The next day, they headed out to Melbourne, and they had not only a puncheon of their collaboratively made experimental wine in the back of the car but also the bounty they'd gathered at James's farm: several wild rabbits and goats and a cooler full of spear-gunned octopus and abalone.

A sizeable group awaited the oddball winemakers at Gertrude Street Enoteca, an Italian-focused wine bar located in the fashionable neighborhood of Collingwood. It was surrounded by low-rise Victorian houses with terracotta roofs and patios distinguished by cast-iron lace. These residences mingled with eclectic graffiti, vintage clothing and vinyl record stores, and plenty of eateries serving both modern Australian food and *báhn mì* sandwiches.

It was summer in Melbourne, and its wine lovers were ready for a party. Word had been getting around in hospitality circles about the blond, wispy-haired Wildman and the gentle-toned, glasses-wearing Tom and their strange wine projects. Tom had lived in Tuscany for several years and returned to produce wine from a family vineyard of mostly Shiraz out in the Barossa. Since 2007, Wildman was producing limited amounts of highly coveted wine, including Pinot and Chardonnay from growers around the Adelaide Hills. Tom and Wildman met at a wedding and became fast friends, bonding over their unusual approach to wine, which back then nobody even had a name for.

The lanky and tall, ever-smiling James was transitioning out of a career as an award-winning sommelier. He'd trained at the pedigreed Magill Estate, a wine-centric restaurant on the property of the historic Penfolds winery estate, in South Australia. James then made a name for himself while helming the list of fine-dining restaurant Auge in Adelaide. Around the country, it was a shock that a sommelier from little old Adelaide, whose dining scene was laughable compared to that of Sydney or Melbourne, was winning awards. But James was intelligent, eloquent, and passionate about his sommelier work. As he was becoming increasingly known, he met Wildman. He visited Wildman's newly purchased farm in Basket Range, where he was making wine using rudimentary machinery in a few ramshackle sheds, and James was smitten by the romance of it all. He enlisted Wildman to custom-make some wines for Auge. Unfortunately, as word got around Australia of James's affiliation with experimental winemakers while in such a visible position, he received verbal insults from the older wine professional establishment. But he persisted, undeterred by the notion of not being mainstream.

A recent trip to Europe, not long before this journey to Melbourne, had inspired James to start making his own wine. He was gaining experience at a winery in the Hills and planning to make his own wine in the coming vintage—Wildman had offered space in his shed to this new friend. These three, along with another South Australian winemaker, Kerri Thompson, had come to Melbourne to release wine that they had made without any commercial yeasts or acids, and low amounts of sulfites. The wines were also unfiltered. It was all very novel and provocative.

Tom, Wildman, James, and Kerri arrived at the swank Gertrude Street Enoteca. Tom had a goat carcass, wrapped in a bedsheet, over his shoulder, and James bore the cooler of fresh seafood. Chef Brigitte Hafner got to work breaking it all down. Meanwhile, the South Australians went around the bar pouring their cloudy wines.

"Really, you didn't add *anything* to this wine?" wine critic Jeni Port questioned Tom Shobbrook.

Tom shrugged and said, "Only a bit of sulfur."

Jeni sniffed and tasted the Shiraz rosé—and frowned. "This wine is a bit brackish and dirty, Tom. I don't know if it quite works." Without any stabilizers or acidifiers and with zero filtration, Tom's nearly naked wine was unpalatable to the critic. She was more accustomed to Australian wines that aspired to the Robert Parker standards of robust, silky textures and strong alcohol and tannin. Those wines were always filtered and often corrected with additives.

"Dirty—that's right! I love it!" hollered Tom. He took a big slug of his own wine, grinned, and went off to pour for other guests.

Jeni approached Wildman, who, with his sinewy eyes and thoughtful, slow tone, she suspected was the philosophical one in the group.

"You must have been inspired by the work of that American journalist, Alice Feiring?" She assumed that Wildman would have read the books of natural wine's most prominent champion. But he shook his head, said he'd never heard of her, and had no idea that people outside Australia were already making wines like these.

"Natural wine isn't a term I'm familiar with," said Wildman. He explained that the project they had in mind was more about freedom, breaking the rules. He'd been making wine without electricity and adorning his bottles with labels made of handmade paper. He had also planted a vineyard of Pinot Noir that he intended to never irrigate and never spray, not even a single time with copper or sulfur, common organic treatments. The idea was to be different, to create something singular.

"Sounds interesting," said Jeni as she tried Wildman's Pinot from another vineyard in the Adelaide Hills. Hiding her sneer, she jotted down in her notebook, "childlike simplicity," "not fully formed," and "unusually light and gentle in the mouth."

A few weeks later her article about the tasting came out in Melbourne's the *Age*, entitled "Natural Selection Theory" in a cheeky nod to Darwin. It praised Kerri's Rieslings and decried Wildman and Tom Shobbrook's wines as dirty and flawed. Kerri soon went her own way, but the others continued forging their bond and pushing the limits of wine. And now they had a name for their experiments.

After the event in Melbourne, the three men drove straight to Sydney, sleeping on the side of the road in rolled-out "swags," the black mattresses that Australians love to bring on road trips. There, they caught up with their friend, the charismatic musician and wine industry professional Sam Hughes. The foursome was soon to be known across Australia and abroad as the Natural Selection Theory collective.

In 2008, Tom and Wildman had encountered Sam while on a sales trip to Sydney. It was the early days of their friendship and the beginning of their careers as independent winemakers. Flipping through the pages of the *Good Food Guide*, Tom and Wildman wound up at a bottle shop in Sydney named Vaucluse after an appellation in France's Rhône Valley. Sam was the shop manager, and upon trying the South Australians' unconventional wines, he was enamored.

Wildman and Tom, who both had young children at home and zero outside investment in their fledging wineries, were traveling on a budget. When they told Sam they planned to sleep in a park that night, as a Sydney hotel was out of their price range, he invited them over for dinner and to stay the night. Back at his place in the neighborhood of Marrickville, where visual artists worked in former industrial warehouses, Sam poured Tom and Wildman a wine that would forever change them: a skin-contact Ribolla Gialla from a pioneering no-sulfite winemaker in Italy's northern region of Friuli—none other than Stanko Radikon.

Tom and James and Wildman continued getting to know Sam in the summer of 2010, during further sales trips. They drank bottles of wine at the chic bar 10 William Street, another early supporter of Tom and Wildman's first vintages. The more they tasted, the more the quartet wanted to break the rules of Australian winemaking.

"I've been reading this book by a Japanese writer called Masaru Emoto," Wildman told them one night at 10 William Street. "It's all about how water responds to energies. He did experiments in which water is exposed to different human emotions and thoughts, such as affection or anger, and the actual molecular structure of the water is altered completely." Wildman's blue eyes burned with passion as he described these trials. "We should try something like that with wine."

That's how the egg project was hatched. James knew of a vineyard in the Hunter Valley region of New South Wales, to Sydney's north, which had been planted by Leonard Evans. Leonard Evans was a well-known "publican," as Australians respectfully call individuals who own pubs, and he was one of Australian wine's early champions.

Sam found a warehouse in Sydney and made some phone calls to procure a few thirty-liter ceramic eggs. With a half ton of Sémillon from the Leonard Evans vineyard, some of which was fermented on skins in separate ceramic eggs, the men set out to attempt to replicate the Emoto water experiments. They played recorded voices repeating, respectively, "love" and "hate," and accompanying music and sound to the wines, and even installed corresponding soil types surrounding the base of the eggs. The experiment went on over the course of a month, to see how the wines turned out differently. They had chosen Sémillon from Hunter Valley precisely because it was a prestigious region's prized variety. As when they'd appropriated Jeni Port's article title as their posse name, they were "taking the piss."

Natural Selection Theory was dismantling everything the Australian wine industry aspired to. It was irreverent and playful, uninterested in Robert Parker scores or coveted awards. It was scrappy and handmade. The egg wines made in Sydney, and the sulfite-free wines that Wildman and James began making in the Adelaide Hills, while Tom began his own attempts in his Barossa Valley winery, were attracting plenty of press and critical acclaim. In part, it was the charisma these three winemakers, along with Sam, radiated, each in different ways. Together, they were an unstoppable party, driving around Australia with their demijohns full of sulfite-free wine, which they called "Voice of the People," and one-liter egg-shaped vessels filled with the experimental "Love-Hate" egg wines. They even recorded a *Natural Selection Theory* album, with each of them playing an instrument or doing vocals in their respective studios, and pressed a few hundred vinyl records. The point wasn't to make money; it was to show that wine didn't have to be snobbish or conceited or capitalistic. It could be uncanny and unpredictable, and the life of the party.

In James's shed in Basket Range, Sev, Alex, Raph, and I stood mesmerized as James and Wildman recounted those early days, a decade ago. But then they both fell quiet and turned their eyes downward. I understood why. I knew how Natural Selection Theory had ended. Sam had died by suicide, very suddenly, at the end of 2012.

Following that tragedy, James and Wildman and Tom each went their own ways, continuing to make natural wine for their own brands. By 2016, all three wineries, Lucy Margaux, Jauma Wines, and Tom Shobbrook Wines, had emerged as entirely no-sulfites-added wineries.

James finally thanked us for coming, and we headed back to the farm with a bottle of his Like Raindrops Grenache. We drank it that night with a dinner of kale salad and roast chicken. The wine was sultry and rich, and to me, it also tasted a bit of sadness, of longing for a person who could never be replaced and would never reappear, someone lost to his own brilliance.

By late April, all the wines were nearly finished with fermentation. Interns from neighboring wineries showed up, and Wildman led us all around his shed in a big cluster so we could taste the product of our labors. As we went along, we reflected on how each decision along the way—when to pick, when to press, blending this with that—had impacted the wines' profiles.

Raphael, Sev, and Alex were preparing to leave Australia and return to Europe. We decided to throw another party, just outside the shed, to bid them farewell.

Two days before the party, I awoke at dawn to find the bed empty beside me. The dogs weren't around, either. I sat up and got out of bed. Outside, behind the house, I looked out over the fog hanging in the valley. Then I heard it—*pow*—the first shot, followed by a second one a few minutes later. I had a pretty good idea what was happening.

I went inside to make coffee. When the fresh brew was ready, I brought it with me outside, just in time to see Wildman riding along the path on his beat-up yellow tractor with a small trailer attached to the back. The sheep's carcass—one from his own flock, which normally roamed on the hillside just beside the house—was splayed across it, and the dogs jogged happily along on either side. Wildman caught my eye as the tractor rolled by. The skin around his eyes crinkled into a smile, but I detected a glimmer of sadness, or perhaps reverence for the act he had just committed.

"Good morning," I called out over the engine, feeling ever so the farmer's wife. My eyes were naturally drawn to the animal, and they registered the sheep's head, twisted to one side, blood oozing out of a slit in the neck. I followed the tractor as it rumbled down to the winery. The others had awakened, probably also having heard the shots, and they emerged in their pajamas at the winery. Nobody wanted to miss this.

Using the same tethers that we used to hitch up the trailer when it was full of heavy bins of grapes, Wildman tied each leg of the sheep to the awning of the winery that hung over the concrete just outside. He didn't speak, didn't look at anyone but the sheep. It was a big one, older. Wildman placed a bucket underneath the animal. Sev and I passed a pouch of tobacco between us, quickly rolling and lighting cigarettes. With a simple pocket knife, starting at the feet, Wildman carved the wool and skin from the sheep, moving down until they were hanging from the sheep's shoulders, covering its face. Blood dripped along the carcass into the bucket.

Little by little, Wildman began opening the sheep's stomach.

With that sheep hanging there, this animal whose presence had been part of my daily life for three months, I felt incredible awe. Watching Wildman ceremoniously removing the innards from the belly was definitely the most intimate I'd ever been with any meat that I'd eaten before.

That night, everyone from the natural wine community around the Hills came to the farm, bearing magnums of their wines, and we feasted on the mutton, along with kale and fennel salad. Alfie and Lulu hung out nearby until we were too drunk to notice them snatching entire legs of meat from the grill. Wildman and I laughed about it and went back to drinking and dancing.

A few days later, the interns had caught their flights to Europe, and the farm was quiet except for the cawing magpies. Sipping my morning coffee, I cuddled with Alfie and Lulu on the dirty couch. It wasn't the only dirty item in that room. Once again, cobwebs had crept into the corners. The tile floor was dusty with footprints. Dishes were piled in the sink. Just about everything in the house needed intense scrubbing and laundering. Wildman seemed preoccupied with paying invoices from the growers, so I started the job. I hoped to have the house somewhat in order before the one-month trip to Europe we were planning.

At the start of May, Aaron came over to the farm to watch over Alfie and Lulu and check on the ferments while Wildman and I were away. I waved at the little green shed, where Persephone lay in rest. There were five wines in barrel in there, and my pét-nat was already bottled, awaiting disgorgement.

We were first headed to Copenhagen, where Wildman had been invited to pour at a festival called Fri Vin. Following that, we visited winemakers in Champagne, Slovakia, and Italy and spent a few days in Paris. There, we tromped around town and spent time with Gaba, who was visibly pregnant. She and Dan were fixing up their new apartment, applying fresh paint and installing a cot and change table, preparing for the baby's arrival in September. Pregnancy and nesting seemed to be exhausting Gaba physically, so we didn't go out on the town much, although we were sure to have lunch at Les Arlots, our eternal favorite, where we'd had that bottle of Lucy Margaux Vino Rosso about two years earlier, though it felt like a lifetime ago.

In between courses, Gaba and I stepped outside for some fresh air.

"Guess it will be a while before my wine bar opens," said Gaba.

"It might be a while, yes, but that doesn't mean it won't happen," I said. I noted how her wording had changed, from "our" wine bar to hers. What could I say? She was right. I'd officially moved away from Paris, and it didn't seem likely that I would return. I was making wine, building a relationship, in a country very far from Europe. After lunch, Gaba and I hugged intensely, and we promised to speak regularly.

"I'll be back no matter what for the baby's birth," I promised. A twenty-four-hour flight wouldn't stop me from returning, even if I had to sleep on her floor and could only spend a week there.

In Paris, we also celebrated the arrival of *Terre* Issue 2 with a small event at La Buvette. Across the pond, my creative partners did not hold any events. I felt a change in our relationship—they didn't appear excited anymore about the magazine. But the days were flying by, and I hardly had time to think about it or speak with them, other than emails regarding the distribution of magazines to our stockists and subscribers.

Wildman and I took a flight to Italy for some time alone. We celebrated my thirty-fourth birthday in a house we rented in the hillsides of Lazio, outside Rome. Wildman cooked steak over a fire he built amongst some rocks, and we ate in silence under an open sky.

We flew from Rome back to Adelaide; I was starting to get used to the long air journey. I used it to think about all the changes in my life. During that month, back in Australia, our wines were slowly coalescing into entirely singular entities, which could never again be repeated. I'd been in touch with Chris Terrell, the importer who had first invited me to Georgia, and he was making plans to import Wildman's wines and mine over to the US. The idea of friends back in the States enjoying my Persephone creations thrilled me.

As we drove into Basket Range, I craned my neck at the towering pine and eucalyptus trees; thick gray clouds started moving in around their canopies. The Land Rover careened up the hill from the veggie patch to the house, where we could hear Alfie and Lulu barking, welcoming us home. As if on cue, the sky opened up. Rain began pouring down, diagonally with the wind.

"Ah, the sideways rain," said Wildman as the car pulled up beside the house. "That means winter is here."

"Ugh, rain," I said. "Not my favorite."

"Well, it's good for all the plants and the vines," was his practical reply.

We unloaded our suitcases quickly and brought them inside.

"I'll go grab some firewood," said Wildman. He headed down to the shed, leaving me alone in the house. I shivered and wrapped my arms around my body. It felt different, this time, arriving at the farm. Cold and rainy, for one thing. And it was just us now. No interns. Just Wildman and me and the dogs, who were soaking wet and staring at me through the glass doors of the kitchen. I went to look for a towel, then let them in and began drying them off, scrubbing their thick coats and rubbing their paws, one by one, until the towel was soaking wet with their filth. And from that moment onward, I was no longer living in a cosmopolitan paradise, with constant social activity and fabulous weather. Now I was in the Hills during the rainy and cold winter, which for me would be a time of loneliness and questioning.

– Thirteen –

Pét-Nat and a Pantry

The beginning of June was very strange. Freshly glowing from the romance of that month spent in Europe, I confronted a bleak truth: the *Terre* magazine project was not going well. Emails and phone calls between my two creative partners and I had become strained and confrontational. After some heated exchanges, we decided to part ways. I couldn't help but attribute fault to the distance between us. During my brief trip to the US four months earlier, we had failed to meet in person because of the blizzard, and subsequently, producing Issue 2 of *Terre* via remote, online conversations had been unfulfilling. While I wanted to make our magazine even more explicitly focused on natural wine, I could see that my two partners were ambivalent—it wasn't their true passion.

I agonized about the loss of this venture. Seeing the physical magazine I'd helped create in the hands of people in New York, Europe, and Australia, I'd felt pride and excitement about the future of this independent print publication. I had put in countless hours editing articles, fussing over design, and emailing contacts to support the magazine. If it disappeared, would I go back to the tenuous existence of writing freelance for various culture websites?

At 3 a.m., while the rain—which hadn't ceased in days—landed softly on the tin roof, I lay awake, replying to emails from my creative partners, desperately trying to salvage things. The distance between us was immense, both geographically and mentally. I felt incredibly alone.

Wildman's hand, warm and strong, crept out from under the covers and onto my arm, squeezing gently.

"I'm sorry, I woke you," I sighed. "It's hard to believe it's ending." I momentarily paused the motions of my fingers on the screen. "I loved making that magazine."

He nodded, rustling his head on the pillow. "And people love reading it. But surely you can keep it going on your own?"

The idea had definitely occurred to me, but it gave me intense anxiety. If I were to continue making a magazine, I would narrow its scope to focus entirely on natural wine. But the skeptical voices of my New York editors, telling me natural wine was too niche, echoed in my mind. How could I be sure people would want to read my publication? And starting anew, with no money, would I even be able to afford several thousand dollars in printing costs? Could presales alone fund that?

Wildman drifted back to sleep, but I remained awake for hours, letting my insomniac mind work through all of these questions. In miles as well as in lifestyle, I couldn't be farther from New York City, yet its problems continued to plague me. Maybe, if I created my very own publication, I could sever my remaining emotional ties to the City. Over the years of working in natural wine I'd met plenty of people, around the world, whom I felt would support me if I struck out on my own.

The next morning, I began sketching out a plan for a rebranded publication. The month Wildman and I had spent in Europe was a source of inspiration. I thought back on a visit we made to a winemaker in Slovakia named Zsolt Sütő, of Strekov 1075 winery.

Around two hours outside of Bratislava, the capital of Slovakia, Wildman and I arrived in a small town, back in mid-May. He was interested in importing the Strekov 1075 wines, to sell by the glass at the Summertown Aristologist. I was happy to tag along, as I'd found the Strekov 1075 wines intriguing back in New York, where they were distributed by Jenny & François.

Wildman had taken a wrong turn as we left Bratislava, and we had a classic argument about navigating. Finally we pulled up the rental car outside the large brick structure that was the Strekov 1075 winery. Zsolt, an imposing man at least six feet in height, greeted us. We spent an hour tasting his wines from the bottle, discussing the unique heritage of that part of Slovakia, which borders on Hungary, and then he proposed a small field trip.

"We will see my vineyards," he said. "And then, I want to take you to meet a friend of mine. He makes natural wines like no other." We agreed happily.

This friend was in a small cave, behind a quaint row of houses not far from the vineyards. The entrance to the cave was nearly covered by wild-growing trees and flowers. A short man with high cheekbones, wearing a pageboy cap with strands of blond hair poking out, came out as we arrived. He wore a clean, white button-down shirt, with the top buttons open and sleeves loosened. Zsolt presented him as his friend Gabor and explained that he spoke Hungarian. We followed Gabor and Zsolt inside the cave. We met Gabor's wife, who was from Japan. Then commenced a tasting like none other we'd experienced before.

The cave held nothing more than a few barrels full of wine. Zsolt, translating, explained that his friends spent half the year working as string musicians in Hamburg. They made this wine completely naturally from organic vineyards, for personal consumption. Rather than putting it into new bottles with attractive labels, they siphoned it directly from the barrels into used French wine bottles and distributed it amongst friends. It was noncommercial natural wine, made of grapes found only in that part of Middle Europe, like Welschriesling and Saint Laurent, and it was delicious.

Near the end of the tasting, we stumbled out of the cave into the fresh air. Gabor periodically ran back in to get more wines for us to try, using a thick-glass *pipette* to draw wine, and then distributed splashes into our glasses. A pipette is an instrument designed to extract tastes of wine from barrels. I knew it by the French pronunciation: "pee-pet." The one Gabor was using now was particularly elegant and beautiful: it had a crescent-shaped handle whose opening was just wide enough to fit one of Gabor's strong fingers, and a graciously round bulb where the liquid gathered.

"I can't believe how lovely this pipette is," I said, slurring my speech. We hadn't eaten since the morning and our tasting had gone on for hours. "Is it antique? It's very special."

Perhaps it was an heirloom pipette—perhaps it was deeply treasured. Nevertheless, as soon as Gabor saw my appreciation of this instrument, he made a decision. Once we'd finished the three-hour-long, deeply pleasurable tasting, he carefully cleaned the pipette with a white cloth and handed it to me. Wildman and I tried to refuse the gift but he wouldn't hear of it.

The generosity struck me. I felt that this ethos was central to natural wine—for most producers, it was ultimately a business, but profit wasn't the only motive. I had a bad feeling, though, about our chances of getting the pipette back to Australia. Sure enough, a few days later, as Wildman and I rushed to catch a flight in Vienna, the pipette shattered.

The ephemerality of this gift brought more reflections on natural wine's essential purpose. After all the work that goes into producing wine, it's consumed in a matter of minutes, but the emotional wellness and community it creates can last much longer.

I thought this French term, *pipette*, spoke to the new magazine's mission to showcase different stories of natural wine producers and industry professionals. To get the ball rolling, I contacted writer friends in Europe and North America and assigned articles for the debut issue of *Pipette Magazine*. I planned a trip to the Australian state of Victoria to write about Momento Mori, an upstart natural wine producer I'd first encountered at Rootstock in Sydney.

Every morning, while Wildman puttered around the property, fixing a tractor wheel or ordering bottles or managing the gardener in the veggie patch, where broccoli and cauliflower were coming up, I sat down with a coffee and emailed people who had supported *Terre*, crossing my fingers that wine shops, bookstores, and other retailers around the globe would be interested in something exclusively focused on natural wine.

I felt a mixture of nervousness and creative inspiration while planning the new publication. It stemmed largely from feeling uprooted. Over the phone, Gaba filled me in on visits with her midwife and translation work she was doing for an importer, and I updated her on farm life. I FaceTimed my siblings and mother to show them this pastoral paradise where I was awakening with the birds, surrounded by valleys where sheep roamed. But I still had no local friends in Australia to whom I could vent about this major rupture in my life, the dissolution of *Terre*. Meanwhile, the days grew shorter, grayer, and wetter. The winter solstice approached. Thankfully, Wildman and I had plenty of work cut out for us, preparing our wines for release that coming spring, starting with the tediousness of disgorging our pét-nats.

PFFFFFFFFFZZZZZZZZZZ. Wildman popped the crown cap off another bottle and put its neck through a hole in the large plastic bin.

Excess wine foamed out, bringing with it chunks of naturally occurring crystalline deposits known as tartrates, and lees made of dead yeasts and grape skins that had accumulated in the bottle while it rested, upside down, over the past few months. Now, the disgorged bottle was only three-fourths full—it surprised me that so much wine was lost during this process.

"It will be worth it," Wildman reassured me. And I agreed—having opened a few non-disgorged sparkling wines before, I could attest that they too often sprayed everywhere. Better to lose the wine here and present customers with a properly finished wine that wouldn't fizz in their face.

Swaying to the beat of "Hot Stuff," partly to stay warm and partly to keep my energy up, I tilted a bottle of the same sparkling wine into the recently disgorged bottles, topping them back up one by one, then capping them manually on a table. That day we'd repeat this with two hundred bottles of my pét-nat—a pink wine I'd made at the start of vintage from blending separately fermented Chardonnay and Pinot Noir.

"Well, I feel stupid," I said.

Wildman looked at me quizzically.

"If you only knew," I said, reaching for another bottle to pop open, "how many explanatory articles I've written that introduce pét-nat as the 'simpler, easier' version of Champagne."

It was true, of course, that the fancy French stuff was made with two fermentations, while pét-nat had only one. But as I was learning, that didn't make the latter any easier, especially given that we did all of the work by hand, rather than with machine-operated riddlers and disgorgers, as many Champagne houses use.

The challenge began with carefully timing the bottling of the wines. Pét-nat is generally made by bottling wine with a tiny amount of residual sugar, so that bubbles occur as a result of fermentation finishing. In my case, I'd been so busy working the basket press and picking in the vineyards that my Pinot and Chardonnay had fermented dry over the course of a month. So we had added some fresh Gamay juice before bottling, to help fermentation restart. Wildman made the calculations carefully, using a textbook from his student days to determine the exact number of liters we should add. For his wines, he'd been more on top of the timing and had bottled them with sweetness. Either method was fine to us—it was pét-nat regardless. Natural wine was a world without definitions, anyway.

For many who adore natural wine, pét-nat is the heart and song of our movement, something we rely on to set a cheery mood amongst friends or kick off a weekend gathering. A chilled, cloudy, low-alcohol bottle of pét-nat is the bad boy in a leather jacket, while a bottle of Champagne is more like a woman in a Chanel suit with pearls. Of course, there are plenty of Champagne grower-producers who actually do farm and vinify entirely by hand and whose wines can display some very unique aspects. They operate quite differently than large Champagne "houses," which purchase fruit from all over the region and more or less use a recipe to make their sparklings. But pét-nat is the natural wine movement's darling because it refuses to be dressed up.

Back in the day, the folklore goes, in sixteenth-century Southern France, the first sparkling wines happened largely by accident when producers in the town of Limoux bottled their young juice before it had finished fermentation. Bottling wine was a rare thing then (it was served directly from the cask). In the cases where a wine wasn't yet dry, its residual sugar might be converted to carbon dioxide—in other words, bubbles—if fermentation restarted once it had been bottled. Although it happened accidentally at first, that approach was adopted as an intentional way of making sparkling wine and became known as *méthode ancestrale*, or the "ancestral method," in contrast to the method developed in Champagne in the nineteenth century. The Champagne method involves bottling still, dry wine with added yeast and sugar to kickstart a secondary fermentation, and aging it extensively in a cellar so that the sparkling wine develops complexity. Today, some wines labeled méthode ancestrale achieve fizziness by arresting fermentation through temperature control or adding sulfur before bottling, when the process is then allowed to resume. Generally, both méthode ancestrale and Champagne wines are disgorged, meaning they are opened to allow for tartrates and yeasts to spill out, before refilling and sealing with a cork.

Pét-nat may or may not be disgorged. If it is disgorged, it's usually not resealed with a cork—a regular crown cap will do. The specific term *pét-nat* and its natural wine ideology came about in the '90s, with credit going specifically to Christian Chaussard, who made wine in Vouvray until he passed away in a tractor accident in 2012. In Vouvray, the main variety is Chenin Blanc, which can produce a sweet wine if harvested late. Sometime in the late '90s, Chaussard bottled one Chenin Blanc wine with a touch of residual sugar, and he either did not add sulfites or added very little to arrest fermentation—therefore, the wine kept going in bottle and became fizzy. Instead of tossing out the wine, which he found quite enjoyable, Chaussard called it a *pétillant-naturel*, a natural sparkling, and referenced the méthode ancestrale as an example of its historical occurrence. Pét-nat, therefore, is a perfectly postmodern thing in that it's made with an irreverent eye to a modern ancestor. It stands apart from méthode ancestrale wines in that no added yeasts or preservatives may be used—it should be 100 percent grapes-only.

Wildman and I disgorged for hours on end before our toes were numb and our hands were soaking with cold wine and pink, foamy lees, and I called it a day. Wildman probably could have continued for another hour—he never seemed to mind the cold.

"Sit, Alfie, sit, Lulu. That's good." Back in the house, I dried off the dogs' muddy paws, shivering while Wildman stacked kindling in the small old-fashioned wood stove in the hallway. It was only 6 p.m., but outside it was dark and misty. Once the fire was going, Wildman came into the front room.

"My love, what would you like for dinner?" He had such a gentlemanly way of asking me what I'd like him to cook. He made me feel doted upon, cared for.

I looked up at my partner from the floor, where I was stroking Alfie's wet nose. The veggie patch was brimming with kale, chard, and butternut squash at the moment, but I couldn't think of anything I wanted for dinner. All I could think was that I'd made some sort of mistake. That I wasn't cut out for the Adelaide Hills winter. And what did that mean for my future with Wildman?

There were involuntary tears in my eyes. My arms ached from pulling down the manual capper onto the disgorged and refilled bottles. I felt deeply chilled in my bones. I hadn't written for days and had not had any kind of meaningful social interaction aside from with Wildman—and the dogs and Aaron and a few others at the restaurant—since returning from Europe.

"What's wrong?" Now his tone had changed to concern.

I stood and looked around at the state of things in that space, which combined a kitchen and living room. There were jars of olives and lacto-fermented wild mushrooms stacked alongside pots and pans on a sad-looking IKEA shelf. Boxes of salt and assorted condiments like fish sauce and vinegar sat amongst plates and mugs. An open bag of flour, freshly milled at the Aristologist, was spilling out onto the floor, which was an atrocity in itself. Our muddy boots sat by the door, with not even a mat underneath them to catch the drips. Did we even own a mop or a broom? The last one had migrated to the winery and never returned.

Something in me was ready to snap.

"You know what I want," I told Wildman, angrily. "I would like to not live in filth." His face dropped. I continued unleashing my pent-up feelings. "There's stuff everywhere! This place is a mess. How am I supposed to get any writing done?" Without waiting for his reply, I poured myself a glass of red wine and took it into the bathroom, where I started to draw myself a bath, the one thing I knew could offer me comfort. But when the bath was about half full, I stuck my hand in to check it.

"What the fuck?" I muttered to myself in disbelief. It was lukewarm. I remembered: there hadn't been enough sun that day to heat the stored water. A hot bath was out of the question. So much for my one comfort.

Defeated, I drained the bath and sat in the hallway, on the hard floor in front of the fireplace, with my wine. Over the gurgle of the water returning into the pipes, to then flow back into the earth surrounding the house, I heard Wildman banging around in the kitchen. He was upset, I could tell. He was trying so hard to make me happy—he had welcomed me into his life and turned me into a winemaker! We'd just spent a month in Europe. I knew I seemed ungrateful. But couldn't he see my side of things? I'd never thought, nearly one year earlier when Wildman and I had been traipsing around Europe, throwing back glasses of cult wine and eating in cozy bistros, that I'd leave my dreams of Paris behind to work long hours on my feet in a shockingly cold winery, live in a house that felt like a ramshackle work shed, and not even be able to take a warming bath.

Wildman came and sat next to me, tossing another log on the fire. I looked at our dirty jeans beside each other, two sets of legs covered in spattered wine lees. My eyes found his face, and its pleading expression.

"I'm sorry," he said. "I want this to be your home, too. For a long time, even though I was married, it was like I lived alone. If I sometimes don't think about your needs, it's because I'm not used to doing that."

I felt calmer and apologized for my outburst. We went back into the kitchen and had dinner of roast vegetables and chicken. But what he'd said didn't quite make sense to me. Hadn't his ex-wife also lived here? How did she manage with the filth? The lack of hot water? But then I began to see—that was why the marriage had crumbled. They ignored the problem instead of dealing with it.

The next day, Wildman disappeared early in the morning. He returned and informed me of two things: one, someone was on the way to install a gas heating system for the bathroom's water supply. And two, he'd procured a ton of bricks. We were to start building our pantry on the weekend, in that miserable empty room beside the kitchen—as soon as Wildman had knocked out the existing wall and broken up the tiles on the floor. Until then: more disgorging.

We did have a few days off from disgorging when the sun shone through the clouds momentarily. On one of those mornings, Wildman and I drove to a nearby vineyard where he'd grafted Trousseau, a red variety from the Jura. From those vines he had made just one vintage of wine, a magical concoction tasting of barely ripe plums, violets, and soft woodsy tannins.

But then he lost his right to harvest the Trousseau.

"After all the work I did grafting this vineyard, now some big corporation is taking it over," Wildman grumbled. Apparently, all of his relationships with growers were handshake deals. We knelt down by one of the vines, and he showed me one of the bare branches growing off its trunk.

"You cut just above the second knob," he said, pointing. "This one is nice and straight, that's a good cutting." He snipped. The plan was to send these cuttings of Trousseau to a nursery, where they'd grow into rootlings over the course of a year. Then, we would plant them on the hillsides at the farm. Using the very same pruning shears that Wildman had gifted me in Paris, I worked beside him, snipping off vine branches and carefully bundling them into piles of one hundred. We were making the future of a wine estate.

It took five days to build our pantry, with me applying the mortar and Wildman doing the meticulous bricklaying using a leveler. I had never done any sort of home project like this. In my family, we had either paid someone to fix things or just left them broken.

When it was all finished, Wildman and I assembled and installed some shelving units and immediately filled them with the kitchen items that had cluttered the living space. The next morning, I felt like I could breathe more easily as we made pour-over coffee, no longer surrounded by jars. For the first time, I was able to sit calmly in my office, writing and looking out the large French windows that opened onto the valley where the sheep liked to graze. I was working on the first issue of *Pipette* and starting to feel more confident about it as the articles took shape and people responded with their preorders and advance payment.

Life on the home front was drastically improved by our pantry, and we moved on to discussions about building a guest bedroom and bathroom for harvest interns, and redoing the entire front room, which had rotting prefab walls on two sides and tile floors that were impossible to clean. But I still craved an existence in the world apart from the farm.

Our local urban center, Adelaide, nicknamed "city of churches" because there are so many, wasn't quite as happening as New York or Paris, but I was determined to explore its cultural offerings. There was one main problem: to get to the city, I needed to learn to drive on the left side of the road, ideally in a manual-shift car, since that was more common in Australia.

Car ownership was a thing of my past—I'd sold my dark green automatic Acura Integra at the age of twenty-four before moving to New York City and had driven rarely since. Wildman gave me a few stick-shift lessons in the Land Rover. White-knuckled, I stalled out on the first big hill leaving Basket Range. But I was determined to learn. Using an app for secondhand goods, we found a cheap car by Holden, an Australian manufacturer, for sale nearby. It was nearly as old as me. I began driving it on short trips, mainly to the post office and back.

One Saturday morning in July, rain was drenching the hills around the farm. I watched it come down through the ceiling-to-floor windows in the front room. Outside, Alfie and Lulu moped about, caught in the waters, accumulating layers of mud that we'd have to towel off when they wanted to come back inside. Wildman was hunched over his laptop, agonizing over various invoices from the vineyard growers. It was time to pay for all the picking and the grapes. In a few days, we'd begin racking and bottling our wines, preparing them for market. Wildman needed the income in order to make those payments.

My mind drifted as I scrolled on my phone, seeing what my friends back in Europe were up to. The World Cup was happening in Russia, and it looked as if France was headed for the win. I pictured myself alongside Gaba in Paris, cheering on my favorite spectator sport at a typical brasserie, sipping cold tap beer. The farm was immensely quiet, and my mood was bored and lonely.

I looked up a dance class online, and made a plan to head into Adelaide that afternoon. That was what I needed: sweating to music and meeting some "city people." A few hours later, I was ambling down Greenhill Road, swerving left and right as the Holden hugged the curves of a ridge overlooking a forest below. Beyond that I could see Adelaide. Its lights were dimly visible in the gentle rain, which covered everything like a sheet. Beyond the city, there was the Great Australian Bight, whose waters were periwinkle blue under the gray winter sky. In the extreme distance, there was Antarctica, where many early South Australian colonizers had made expeditions, often fatal ones.

The dance studio was in Adelaide's city center. Once I found it, though, I realized I hadn't anticipated how difficult it would be to find a parking spot, especially one that could fit a sizeable car. Twenty minutes of searching was fruitless, and the class was about to begin, so I pulled into a garage without reading the signs and ran frantically to find the studio. I got there five minutes late. It was a hip-hop class, and I smiled with a touch of pride as the young instructor gave homage to "Brooklyn, the home of hip-hop."

My spirits were lifted by the endorphins my body produced, and after class I thanked the instructor and said I would return soon. I found my way to my car and maneuvered with difficulty out of the garage, stopping to pay.

"What the . . . " I couldn't believe it: the garage charged me twenty-one dollars for the seventy-five minutes I'd been there—more than the class itself had cost.

Somewhat annoyed, I began the ascent into the Hills, using my phone's GPS to guide me. The app suggested a different route than the one I'd taken to get down, and this one turned out to be equally hilly and windy. As I navigated the car, I began to sweat even more profusely than I had while dancing. It would have been difficult driving on a normal day, but now the evening fog hung low, and I could barely see one meter in front of the car. I went slowly. Hardly anyone was on the road.

Ten minutes later, I was lost. The app seemed to be misdirecting me, and my nerves were shot from the intensity of driving. I pulled over, turned off the car, and let the tears flow. Why had I left Paris, its conveniences and easy living, and the comfort of friendship with another American to live in these damned hills? I hated driving and could never see it becoming an enjoyable part of daily life for me. But without driving, my life would be limited to winery and farm work, housekeeping, cooking with Wildman. In New York, I went out to a new restaurant or a stimulating wine tasting at least twice a week. Dance and yoga classes in Brooklyn and Paris tamed my fiery energy so I could sit and write.

Eventually, I found my way back to the farm, but my spirit was somewhat broken. It took weeks before I felt comfortable driving, but even then, I resisted opportunities to go out. I didn't return to dance class. Instead, I began fantasizing about getting back to Paris and resuming my life there.

Bottling took about two weeks. We started with Wildman's wines and then did mine, which consisted of five reds. By the time we were on to my Cabernet Franc, I was fairly used to the sequence: we first racked the wines from barrel to a stainless-steel tank, which involved dipping a gas-powered siphon into the barrel, careful that it did not capture the lees as we siphoned the liquid. Racking, to me, was incredibly difficult work, as it involved precision—dipping the siphon just to the right level, finding where the lees were, and stopping the gas flow just as the wine ended and lees began.

Once the wine had been racked, we hoisted the tank up above the bottling line, a state-of-the-art machine Wildman had recently purchased from an Italian manufacturer. One of us would load empty bottles onto the line, while the other removed them, filled, at the other end. Aaron came by to help when he wasn't busy at the restaurant, as he also had wines to bottle. We wore many layers, and at the end of the day we'd run through the misty rain up to the house, where we would light a fire, open a bottle of wine, and decide what to make for dinner. Generally, we were in bed by 9 p.m.

I called Gaba once we finally finished all that bottling.

"When are you coming *back* here?" She wanted to know. I pictured her standing outside a tabac, trying to decide where to have her morning coffee and croissant.

"Well," I told her. "In fact, I have good news." I'd been invited to visit Slovenia in mid-August to write about natural wines there. From Slovenia I would make my way to Paris. Gaba was due to give birth in September, and I planned to stay there until she had the baby, to lend support.

"I'll bring a bottle of my wine for you," I promised Gaba. What an amazing thing to be saying, not even a year after my harvest experience in France, when I'd made a fool of myself in the Mosse family winery.

Gaba was elated. "I've missed you so much. I can't wait to see you." It had been a long time since someone had said something so warm and caring to me—besides Wildman, of course, who said often with words as well as his body that he loved me. But speaking to Gaba reminded me that I required more from life than intimacy with one partner.

Australia is nearly as large as the US, although it's less populous thanks to large swaths of harsh desert. It is divided into six states, each of which has its own regional "personality," with its own Indigenous and colonizer histories, capital city, microclimates, natural wonders, accents, culinary specialties, and so on. South Australia is known as culturally conservative. The state was founded by wealthy, land-owning English settlers, as opposed to most of Australia, which was a penal colony.

The original inhabitants of mainland Australia, a complex network of cultures collectively called Aboriginals, are considered the world's oldest living culture. Australian Indigenous culture includes Torres Strait Islanders, from a collection of islands off the coast of Queensland, who are ethnically distinct from Aboriginals. There is evidence that Aboriginals came to mainland Australia around 50,000 years ago. Despite this, when settlers arrived from Britain in the eighteenth century, at the height of the empire, they declared Australia *terra nullius*, land without a sovereign. Relying on Darwinist thought, the colonizers stripped Australia's natives of their humanity as a justification for murdering them and occupying their lands. The legal fiction and racist policy of *terra nullius* was finally overturned in Australian court in 1977. It wasn't until 1992 that Aboriginal people began reclaiming their traditional lands ("Country") by proving that they had continuously and exclusively lived there, starting with the *Mabo v. Queensland* case, in which Eddie Mabo established the principle of "native title" by arguing successfully for the Meriam people's possession of Mer Island.

As Wildman and I prepared for a visit to the island state of Tasmania during the heart of winter for a natural wine event, we watched a series called *First Australians*, which devoted an episode to the brutal history of Tasmania. Since arriving in Australia, many people had gushed to me about Tasmania's unkempt nature, including shimmering blue lakes, lush forests, and rugged coastline. But I had yet to learn the terrible founding story of the "antipode of the antipodes."

Before the British arrived in 1803, Tasmania was inhabited by as many as 15,000 Aboriginals of the Palawa group. The Brits waged a full-on massacre, in three decades, killing all but around four hundred Palawas. The prominent preacher George Augustus Robinson, acting on behalf of the colony, persuaded the remaining Palawas that they would be protected if they surrendered and even promised their land would be returned to them. Instead, he transported nearly all of them to a remote island off the Tasmanian coast, where they lived in a disease-ridden concentration camp.

It is not an isolated incident of shocking abuse in Australia's history of settlement. In Wiradjuri journalist Stan Grant's book *Talking to My Country*, I'd learned how white people poisoned water holes in Aboriginal communities and about the "Stolen Generation," the thousands of Aboriginal and Torres Strait Islander children who were forcibly removed from their parents' care between 1905 and 1967 and even later, by some accounts. I also read Koori writer Bruce Pascoe's *Dark Emu*, which counters notions that Aboriginals had no agriculture. My new home seemed to be just emerging from what Australians call the "great silence," the time when these past injustices were hidden or denied.

Wildman and I flew to Tasmania in between disgorging and bottling our 2018 wines. Hobart is surprisingly happening despite being a small town: one of the world's most cutting-edge contemporary art museums is there, the town hosts an annual art festival called Dark Mofo, and it is home to a handful of striking, chef-led restaurants, most of which serve natural wine. The restaurant Franklin was hosting a tasting called Bottle Tops, and Wildman poured his wines alongside natural-leaning producers from all over Australia, some of whom were only on their second or third vintage.

The day after the tasting, Wildman and I directed our rental car toward Tasmania's southern coast, where we'd rented a house for two nights. On the way out of Hobart, we harvested some nasturtiums from a sidewalk and plucked a few small lemons from someone's front yard.

Our cabin was situated along a long stretch of bay water. It was littered with oysters. Within minutes of dumping our bags, Wildman and I rolled up our jeans and waded into the water, him bearing a small knife, and me holding a lemon half. We gathered roughly a dozen bivalves from the shallow water, some as big as the palm of my hand, and carried them in Wildman's T-shirt up to the house. We sat in front of a crackling fire, opened a bottle of "Wildman Blanc" leftover from the tasting, and savored the freshest, saltiest oysters I've ever had.

The next morning, I woke before Wildman and looked out toward the bay. We were at the end of the end of the world.

I wanted to feel the romance of the moment, but instead, I felt incredibly lonely.

As Wildman and I made pour-over coffee and ate toast, I masked my melancholy. We spent the day hiking through a eucalyptus forest, where we spotted a pair of potoroos—a smaller relative of kangaroos—mating frantically, which gave us a good laugh. We got caught in the rain, a perfect opportunity to drink more wine and eat another handful of oysters under the cover of the canopy.

Back at the house, Wildman refreshed the fire and I changed into sweatpants.

"How about a movie and some snacks?" Wildman suggested. He propped up his iPad on a coffee table, and I prepared guacamole and threw some tortilla chips into a bowl. We had started watching *Barbarella*, a hilariously dated, cult science-fiction movie, the night before. But as Wildman sat down, he paused before restarting the film. I felt his eyes upon me and turned to meet his gaze.

"So, I know we've talked about it, sort of," he began. "And, well . . . will you?"

I looked at him, waiting for him to continue. He couldn't possibly be asking?

"Will you marry me?"

Was he really asking me to marry him, right in front of the iPad, with Jane Fonda on the screen, me in my cheap cotton sweatpants and thick wool socks, and a pile of guacamole in front of us?

When we returned to the farm in Basket Range after our short trip to Tasmania, we were technically engaged. But I had no ring on my finger, we had shared our news with no one, and I felt strange about how it had all happened. Couldn't Wildman have picked one of the *romantic* moments, when we were eating our freshly harvested oysters and drinking wine in the rain perhaps, to propose?

But then I stumbled upon a jeweler and natural wine lover in Melbourne who had ordered a copy of *Terre* online. It seemed like kismet. I commissioned a ring made of rose gold and Australian blue sapphire, and Wildman sent a deposit.

We resumed our routine of farm and winery work, magazine editing, hanging out with other local natural winemakers, dining at the Aristologist, and taking long walks with the dogs. Wildman was beginning to label his wines, and it was painstaking—he first drew each label by hand and added text on his computer, printed the labels on a home printer, cut them using an X-ACTO knife, and then, one by one, dragged each label through a glue machine and pasted it onto a bottle. Since I didn't know the first thing about finding a company to design and print labels, I planned to use Wildman's tedious hand-labeling system for Persephone. But I didn't want to copy him exactly, so instead of using his fancy Japanese paper, I drove to an art store to pick out some watercolor sketchpads.

Wildman and I labeled for many hours each day, and in the evenings I worked on *Pipette*. When there was a pause in the constant rain, we foraged under the massive old pine trees that towered above the Pinot vineyard and collected dozens of orange-hued edible saffron cap mushrooms, which we sautéed with butter and added to pasta.

I made a trip over to the state of Victoria to visit Momento Mori, a budding natural winery focused on Italian varieties and skin-contact whites run by Dane and Hannah Johns, a couple about my age who had learned about farming and winemaking entirely on their own. Hannah was heavily pregnant when I visited, and her cheeks glowed as she threw grain into the paddock where they kept ducks, used to keep snails out of the vineyard. After, I stopped in Melbourne and checked out a few of its acclaimed natural wine bars. It comforted me to know that this vibrant city was only a seventy-minute flight from Adelaide.

Wildman greeted me with a warm hug and kiss as if I'd been gone for weeks. He had seen Lucy while I was away.

"Did you tell her?" I asked. I'd only mentioned to my mother that we were engaged. We hadn't really made a plan for sharing the news.

"Not yet," he said slowly, then adding by way of explanation that Lucy was adjusting to the new school she was attending, where the schoolwork was very demanding. He didn't want to give her more to think about.

I was glad to hear it. Because although I said nothing to Wildman, I had hesitations. Partly, I thought we could have had a slightly better proposal scenario. But also, marriage sounded like something from another planet, not the one I lived on. It terrified me. My parents had divorced when I was nine years old and never remarried. I had not even lived with a man until Wildman.

Nevertheless, for years I had known that I wanted to have a family one day. It was clear that Wildman saw me as a woman he could build a life with. With Lucy he was an incredibly caring father, helping her with math homework or asking about her social life. With our new pantry, the house was feeling much better, and we'd been discussing further improvements we planned to make. And so, when I sat at the desk in the room we made into my office and wrote my article about Momento Mori while looking out over the hills, I had the uncanny feeling that, one day, there would be a baby just beside me, in a crib in that very room.

This vision should have filled me with optimism and excitement. But as winter's rain plodded on, I continued to feel a pit of doubt in my stomach. I really missed urban living, and doubted whether winemaking was the life for me.

One night, soon after we returned from Tasmania, our friend Sarah had a party. It was freezing cold, and everyone stood in her backyard near a fire pit, wearing thick coats and wool beanies and drinking beer or whatever bottle of wine was being passed around. Despite the cold, people were content to be chatting and catching up over drinks. Most of the winemakers had been spending long days pruning their vines back, as is customary in winter, and the socializing was welcome.

"Who wants to chop some more firewood?" Sarah's voice rang out over the clamor. I looked at Wildman, who was busy smoking, sharing a pack of rollies with Aaron.

Beside me, a woman named Rainbo volunteered. She and her husband, Gareth, lived in Basket Range with their son, and they made the wine Gentle Folk. Rainbo had the complexion of an angel and clear blue eyes. I followed her to the side of the house, to see if I could help. But then I stood back and watched—Rainbo heaved the axe up behind her head and slammed it down on a large hunk of wood, atop a chopping block. The wood split perfectly in two. A few women had crowded around, and they all cheered. I joined in, but I felt pathetic. I would never learn how to chop wood like that. At home, if Wildman wasn't around to build a fire, I tried to get it going, but I just wasn't skilled. I wasn't a Hills woman. None of this was in my nature.

The feeling that I didn't fit in was compounded by an overall sense of loneliness. I hadn't gotten over the shock of *Terre* dissolving back in June. My best friend in the world was in Paris, about to have a baby.

On a warm day, Wildman and I took a walk on the beach, and I explained how I was feeling. It was an uneasy, sad conversation. By the time mid-August rolled around, I had canceled the ring, and Wildman's deposit had been refunded. A friend in Paris had told me her apartment was available; she could stay with her boyfriend for a time. I boarded a plane to Slovenia for a press tour of wineries there, with a follow-up flight to Paris. I had not scheduled a return.

– Fourteen –

Return to Belleville

Some women need a room of one's own. I, for whatever reason, seemed to need an entire city of my own. A city in which to let my thoughts flow freely, in which to write without distraction. In which to walk aimlessly to the rhythm of my heels click-click-clicking on the side-walk. In which to feel calm in my own emotional restlessness. Maybe I would be walking only in circles, not really heading anywhere. But that is precisely what I required.

I awoke in the eighth-floor studio apartment that my friend had graciously sublet to me. Standing in the kitchenette by the curtain-less window, I inhaled the scent of boiling water hitting ground coffee in a French press while listening to lively chatter from the Senegalese mosque below. The rooftops of Paris glistened in the autumn sunlight. On this clear weekday morning, even the cap of Sacré-Cœur all the way up in Montmartre was visible. Too lazy to walk all the way down the many flights of stairs and long blocks to the nearest bakery, I en-joyed the coffee on its own. Then I sat down to finish a draft of my article about Slovenian wine, slated for a future issue of *Pipette*. The first edition had been successfully printed. In a week or two, it would arrive in Paris—I couldn't wait to hold it in my hands.

I wrote in my underwear because, why not? Nobody could see me—except the teenagers who were sharing a cigarette out the window of the building directly across, and they surely didn't care. Memories of those vineyards—and the strongly tannic, orange wines made of the local Rebula grape—flooded my brain for a blissful hour.

When I needed a break, I stood and stretched, moving around the small apartment. My diary, with my most personal thoughts, lay open on the bedside table, flaunting itself. Dishes sat, piled up in the sink from the previous nights' dinners, but I left them there defiantly. Nobody to impress but myself.

At midday, I zipped up my A.P.C. Chelsea boots and walked carefully down the stairs. Outside, I headed to the next block over and slid into my favorite seat in the noodle shop. I ordered a *bò bún* and pulled a book out of my purse. The other day, Gaba and I had walked across the river to Shakespeare and Co. and withstood the lines of tourists, so now I had a copy of the latest Houellebecq in English. The cold vermicelli, fried pork rolls, sliced raw carrots, and salty beef, all mingled with lemongrass and cilantro, satisfied me immensely. And it cost only ten euro. After lunch, I headed back up to the apartment to edit an article on Georgian wine.

By early evening, I was ready to shut the laptop. I put on a leather jacket and thin scarf and walked back down the stairs and over to the Belleville entrance to the Métro line. Outside the entrance to the subway, the ladies of the night had just arrived to begin their shift. They prowled the sidewalk outside the large pan-Asian grocery store in black fishnet tights, with long, black hair hanging over their shoulders. I saluted their toughness mentally before going under, then half an hour later I emerged in the 11th arrondissement, where it was quiet—the residents of this neighborhood were still finishing up the workday.

Before arriving at Gaba's apartment, I had one stop to make. I pushed open the old wooden doors, painted bright green, of the Septime La Cave.

"Bonjour," the man behind the counter called out. He was alone in the shop, busy stocking a tall fridge with a wide array of fermented grape temptations. I gestured at a newly arrived Nerello Mascalese from Frank Cornelissen, a Belgian making wine on Sicily's Mount Etna. Something special to drink, even though it was a completely normal day, because every single day in Paris was a moment that might never exist again. Because everything was rapidly changing, in my life, in Gaba's life. I walked out, swinging the bottle in a brown bag.

Through the glass door of the ground-floor apartment, I glimpsed Gaba on the couch beside Dan. She'd been growing out her light brown pixie cut so it was now a bob, with fringe threatening to cover her eyes. Dan's beard and round eyeglasses were fixtures of his friendly countenance. I tapped on the glass and Gaba sprang up, then grimaced, placing one hand on her lower back, the other on her protruding belly. Dan opened the door for me. I hugged them both warmly. Any day now, their little one would arrive.

"Can we get out for a bit?" Gaba definitely seemed to have cabin fever. She reached for her beige overcoat. Dan waved us along, saying he wanted to catch up on some emails. He'd taken paternity leave from his job early just in case things started unexpectedly.

We set out toward the Marché d'Aligre. There, we entered the hubbub of the stalls, where vendors called out their offerings. We began filling a tote bag with fresh handmade pasta, the last of the sun-ripened tomatoes, and large, purple bulbs of organic garlic. We chatted as we walked. Gaba filled me in on her latest visit to the midwife, and I told her about the surprising demand for the first issue of *Pipette*, which was in preorder, and about the launch party I was organizing at a bistro in Belleville the following month. I also updated her about Wildman. Pretty much every day we discussed him. Ever since I'd arrived in Paris a few weeks ago, he had been politely messaging me from Australia, where he was still labeling the year's wines, one hand-cut label at a time, in that cold winery.

"Sounds like he misses you," Gaba said, walking with one hand on her lower back. "And you should see how your face lights up when you talk about him."

I swung the tote bag and thought about Gaba's observation. Was she right? I was thoroughly enjoying my independence. For the most part. I did crave Wildman's arms around me at night. I thought often of days on the farm, back in the warmer season, and the way the sun kissed those majestic hills at the golden hour. I remembered the veggies we picked and ate from the patch, the pleasure of having our wines fermenting so close to where we lived.

But I was in no rush to decide anything about my life—I just wanted to be there, in Paris, writing, spending time with Gaba.

Back at her apartment, we savored the Cornelissen Nerello Mascalese, its smoky volcanic tones and red cherry ripeness, with our simple pasta meal. It was a perfect night, in every way.

Later, in Belleville, before drifting off, I texted Wildman, "Goodnight. Sleep well." And then, I added, "Love you."

When I awoke, his reply was there: "Love you too, beautiful. Miss you heaps."

A few days later, the text came from Gaba: they were headed to the hospital. I stood up and shrieked, springing to action. On the way to the Métro, I deviated to the nearby boulevard, where a row of Algerian businesses offered plates of couscous and sticky sweets. I arrived at the hospital bearing a box of baked goods. Gaba was in a paper gown, sitting on the bed. Dan was beside her with a concerned, bewildered expression. He and I each held one of Gaba's hands while she endured a contraction. After a short visit, I left them to experience the arrival of their child on their own.

That afternoon I went jogging, breathily, not unaware that my tendency to enjoy hand-rolled cigarettes in the evening was affecting my stride, in the hilly Parc des Buttes Chaumont. I walked back to the apartment through Belleville, passing by independent bookstores, cult-favorite Chinese and Thai restaurants, dive bars, and pan-Asian grocery stores—all of which made the neighborhood so vibrant and livable. Back in the apartment, I thought about going out. But I didn't feel up to it and instead began slicing cubes of tofu for my dinner.

I poured myself a glass of Alsatian Riesling. It was strange and I was wary to admit it, but I was starting to feel lonely in Paris. Also, I felt tired. For weeks, I'd been going out drinking with new friends—other Americans, mostly. We had fun, bopping from one wine bar to the next. But going to bed alone felt very odd. And my stomach had been hurting, badly, on some days. Maybe my body was in shock, after months of eating nearly exclusively from the veggie patch on the farm. There was surely plenty of MSG in all those cheap noodle dishes I was eating around Belleville. Maybe I was drinking too much natural wine.

And there was something else—I'd almost forgotten it existed over the past year and a half, since I'd been away from my own country.

Anxiety. It was back, my old friend from New York.

Several times in the past month, I had awoken around 2 a.m. with a twinge of pressure between my brows. Midafternoon, just after lunch, I found myself grinding my teeth and involuntarily biting my lip, and my hands started itching as my eczema flared. I willed it to leave me alone—*Listen, I'm in my happy place now, OK? I'm good. You're not supposed to be here.* I tried to alternately drink and exercise it away—destroying my hungover body with a morning session of Dynamo, the Parisian version of SoulCycle, rapidly spinning while the instructor hollered out French hip-hop lyrics.

That night, I wrote in my journal about all the potential causes of my anxiety—the confusion of love, the excitement of my magazine, wanting to be in Paris, and simultaneously missing the farm. I thought fondly of my psychoanalyst in Brooklyn, the bearded one with bookshelves heavy with the entire works of Lacan, who'd smirked when I said that I thought leaving New York would put an end to my neurotic tendencies. As I poured myself the last glass of the fruity white wine, I lay down in bed and flipped through photos on my phone: Wildman, concentrating as he dipped a glass into a tank of fermenting grapes. Wildman, standing alongside Aaron and Jasper at the restaurant, smoking cigarettes and drinking Beaujolais while around them diners enjoyed plates of food grown on the Lucy Margaux Farm. Wildman, shirtless on the patio behind the house, holding Alfie and Lulu as tiny puppies, back in December. Wildman, shirtless in bed, just after dawn, right after we had made love.

I returned to the hospital and held Gaba and Dan's baby in my arms.

"Man, that epidural really fucking hurt!" Gaba said. I laughed—she acted like the whole event of giving birth had been no big deal. I marveled at Rory's small eyes and ears, his fuzzy head. I also marveled at how Gaba's life was taking a surprising turn—we'd come to Paris with the goal of starting a wine bar, with very little clue of how to actually do that. Now she was taking on the role of motherhood. We'd both been sidetracked, and while I hoped that Gaba would one day get back to the original goal—she was a talented and knowledgeable bartender and sommelier—I had to admit, making wine had transformed me, and I no longer felt content just being a consumer. The life of a producer enticed me.

For the next few weeks, I traveled around Europe to research stories for forthcoming issues of *Pipette*—visiting a winemaker in Galicia, hosting a pop-up in Madrid, touring London to write a guide to the city's natural wine bars, showing the magazine at a trendy food event in Barcelona. In each place, I connected with like-minded people who admired where I was headed with the magazine, but the trips took a surprising toll on me. Returning to Paris from Barcelona on a budget airline, which had forced me to repurchase a flight and wait six hours in the airport after I'd mistakenly stashed my passport in my checked luggage, I felt demolished, emotionally and physically. I declared an end to these trips, reminding myself that I'd come to Paris to write, to reconnect with Gaba and support her in the early days of her motherhood, and to check in with myself.

I was relieved, in fact, that Gaba and Dan's son had arrived, because it meant I had something else to tend to now besides my own tumult.

When Rory was three weeks old and Dan had returned to his full-time job, I sent Gaba to the local *hammam* to sweat for an hour. Meanwhile, I walked around her apartment sipping a glass of Chenin Blanc and holding Rory in my arms. He cooed and blinked. We got into bed and I snapped a selfie, which I posted on Instagram, knowing Wildman would see it. Rory smelled like freshly churned butter and wildflowers. I whispered to him, "I'm your auntie, little guy."

In a short time, Wildman would be arriving in Paris. He'd informed me of this trip casually, over the phone—needing to visit this importer, do a sales trip to that city, stop by this person's vineyard to chat about such-and-such, and of course, spend time in Paris. As if the trip wasn't really to see me.

Gaba returned from the hammam and took Rory into her arms as she sank into the couch.

"Thank you so much," she said, gazing down at her son. "I really needed that." She lifted up her shirt and guided the baby to begin nursing.

"How's that going, by the way?" I had no idea what breastfeeding was like, but I imagined it could hurt or feel strange.

Gaba laughed, but Rory didn't notice. "Boring! I'll be here for probably an hour. I'm so tired of looking at my phone."

It was on the tip of my tongue, the news about Wildman coming to visit. But Gaba looked so relaxed and involved with Rory that instead I told her I'd return the next day and to text me if she needed anything. As the heavy wooden doors of the old building closed behind me, I began walking toward the Métro. But I kept going, past the entrance, and continued through the busy Bastille district, toward the river. With a glance at the stately former mansions alongside the Seine, with their signature tall windows, I crossed the bridge. By then, I knew the way to Shakespeare and Co. bookshop by heart.

Joggers and walkers passed me on the bridge. There were clichés about the French being mean and grumpy, but the people I saw in Paris appeared calm and well. How could they not be, living in a virtual museum of European cultural artifacts, speaking one of the world's most seductive and complex languages, enjoying wine at lunch and dinner?

My shoulders sagged with the realization that the visa I'd worked so hard for, the plans I'd made with Gaba, the years I'd spent learning French—it no longer felt like the obvious direction my life was taking. I'd come to Paris thinking I would shake Wildman and forget the farm, the winemaking, our pastoral life with the veggie patch and the dogs, every night gazing at a sky full of stars. Instead, I found myself overwhelmed by loneliness and anxiety, dreaming of the farm: harvesting the olives, their canopies dancing in the sunlight, waking up to fog lifting over the hills.

The following afternoon, I arrived at Gaba's apartment with a small paperback collection of Borges stories.

"Oh, this is perfect! Let's get out of here, I've been inside all day." She placed the book beside her rocking chair.

We packed up the stroller with diapers and a bottle and headed out onto the street, zipping up our leather jackets, with no plans.

"I bet Le Baron Rouge is just opening," I proposed. Gaba nodded with enthusiasm. She pushed the stroller along and said, dramatically, that it felt *so good* to walk. Her eyes had enormous black circles under them, but I thought she looked gorgeous. Walking with her baby made Gaba seem confident and powerful.

We loved Le Baron Rouge because it wasn't a "new wave" natural wine bar like Septime La Cave. There was no sophisticated branding—instead, a bright red awning adorned with the bar's name in Comic Sans font hung above the large windows and the always-open door. Outside, a prim, aproned man carefully shucked small oysters from Normandy and served them with slices of lemon. Inside, it was chaotic and cluttered. The bar was teeming with people standing over bags full of produce from the nearby market. They were loudly enjoying glasses of wine, poured directly from barrels—a throwback to the days of affordable bulk wine, except the quality now was much improved.

We divided up responsibilities—I hustled my way to the bar and ordered two glasses of organic Sauvignon Blanc while Gaba procured oysters. I saw her, through the window, rhythmically pushing the stroller in an attempt to get Rory to calm down, and momentarily it all felt very surreal. Wasn't it only one year earlier, we were sobbing in my bathtub and smoking out the window underneath the freeway in Brooklyn?

We stood outside because there was no room for the stroller inside. "Ohhhhh, il est sage," cooed a woman, standing over Rory. *Sage* is a French word that translates literally as "wise" but also refers to a well-behaved child. Gaba blushed with pride. So did I, feeling like a protective co-mother.

Ushering a freshly shucked oyster into her mouth, Gaba gasped. "My first one in *nine* months!" I applauded her and we clinked glasses.

"Congratulations," I said, gesturing to Rory. "He's so beautiful. And . . . " I took a drink. "I have something else to cheers to. It's Wildman. He's coming in two weeks."

Gaba knew, of course, that I'd been engaged to Wildman and that I'd subsequently called it off. But she didn't know that recently, my conversations with him had taken on a renewed romantic tone. Even while he and I both acted calm about the upcoming trip, it felt undeniably like a reunion.

"I can't wait to see him," said Gaba. "You'll have to come over for dinner, of course."

"Absolutely," I said. Wildman had mentioned doing some travel together in Italy, but nothing was quite confirmed yet, I told Gaba.

"I think he'll be here, or at least in Europe, for about three weeks," I added. "Then I have to go to Switzerland to do some events for the magazine."

She nodded, patting sleeping Rory on the belly. "Should we get another round?"

"Of oysters or wine?" I smiled.

"Both!"

There was more I needed to tell Gaba. That morning, I'd purchased a flight to Washington, DC. Thanksgiving was the one time each year when the entire Signer family—my three siblings and their families, our parents, and some extended family—all gathered, and I felt it was important to catch up with them. Last night, speaking with Wildman on the phone, I had mentioned the trip and said that perhaps he'd want to join me there, so I could introduce him to my family.

I was just setting down our second round of wine when Rory started fussing. Gaba immediately reached for him, lifting him out of the stroller. He looked up at her and murmured as she wrapped a blanket tightly around him.

"You're so lucky, Gaba," I blurted out. "Rory is going to grow up here." I gestured to the Parisians around us, chatting in happy tones, wearing their good autumn coats, with the collars turned up over their silk neck scarves, and their well-worn leather boots.

"Don't worry, hon, once we get the wine bar plan together, you can come back here and do another short-term rental so we can get it opened." There was impressive determination in Gaba's voice. Or maybe it was denial.

I gave Rory his bottle while Gaba stepped into the street to have a cigarette. He gazed up at me, then his eyelids lowered, and he turned his head to the side, drifting back into the world of baby sleep. It was the most peaceful thing I'd ever seen.

There was one place in Paris where I could always go to feel truly better. And it wasn't a wine bar.

Every Saturday, for nearly two months, I suited up in leggings, a tank top, and running shoes, and rode the Métro a few stops to the Marais. I emerged in the shadow of the red-toned scaffolding outside Renzo Piano's Parisian landmark, the Centre Pompidou. Ducking through a few alleys, I came to an old building in the middle of a long street littered with vintage shops and cafés, the Rue de Temple. I then walked onto a cobblestone patio under an open sky, surrounded on all sides by buildings, from which I could hear the most exquisite cacophony: on the ground floor, a piano; one floor up, a tambourine; somewhere in the midst, a teacher's voice calling out, "Plié, arabesque." It brought me an immediate sense of calm. In Brooklyn, over the years, I'd taken Caribbean and West African dance classes. But in Paris, I embraced a new style.

At the Centre de Danse du Marais, I attended a weekly intermediate hip-hop class taught by a fierce, gorgeous instructor, a woman with long dark hair who stormed around the room and once forbade us from coming unless we wore a hoodie and baseball cap. No matter how hungover I was, how much my stomach gurgled from overconsuming acidic natural wine the night before, I made it to that class because I felt myself pushed to the limits in that space. Many of the dancers were obviously professionals. There were stunning Parisian teenagers wearing expressions of staid sophistication and dancing with their hair down in jeans—they didn't quite get the hip-hop attitude right, but they never missed a beat or a step. The instructor raised eyebrows at me on my first day, but then she said, "À la prochaine," after class, and I took it as an invitation to return. I felt it was a privilege to dance in that space, with that instructor.

At the class just before Wildman was to arrive, I made sure to show up hydrated and well slept. I'd been coming for about seven weeks, and that day, the instructor bumped me from the base-level group—the stragglers who always did the last shift moving across the floor doing the choreography—to one just above them. As I twisted and shook and even whipped my head around so my hair blew in a circle (a much harder move than you'd think), I felt modest pride. I was far from the most elegant dancer, but I also was no longer the very worst! *This* was what I loved about Paris—it offered the space in which to expand as a person.

When I think now of that time in Paris, I hardly remember most of the wine I drank. I know that I got a lot of natural wine journalism done, and that I made great strides in producing *Pipette* while staying in that little apartment high up above the lively streets of Belleville, and I connected and collaborated with a few amazing people. Drinks with Gaba at the dive bars in the neighborhood, including a very late pastis-infused night at a shithole called Zorba the Greek, were irreplicable moments. But what I remember most of all, what I'll never forget, what I wouldn't trade for anything, is being right there in Paris for Rory's birth and dancing every Saturday in the Marais up until Wildman's arrival changed everything.

It was the second time in our short relationship history that Wildman was coming to see me in Paris. But the newness of our togetherness, which had so excited me just over one year earlier, had been replaced by a feeling that we were building *something*, and we needed to figure out what, exactly, that was.

He arrived in the morning, huffing and puffing from walking up the eight flights of stairs, and fell into my arms, sweaty and more muscled than I remembered. There is a lot you can tell from an embrace with a lover you have not seen in a while, and ours announced that this visit was going to be heavy with emotions.

"Well, we have plans tonight," said Wildman, drinking a glass of water in the kitchen, sitting at the table where I'd been writing, looking at the view, which had been *my* private view for so many weeks.

"Oh, do we?" He had mentioned over the phone that I should set aside a nice dress for the evening.

"The operaaaaa!" Wildman couldn't contain the surprise any longer. He announced it with extra enunciation, emphasizing the decadence that awaited us. He had always talked about his love for opera and wanting to take me.

A burgundy American Apparel dress that hit me mid-thigh, and black Chelsea boots I'd purchased on the fashionable Rue de Charonne, near Gaba's apartment, was the nicest outfit I could muster. In the early evening, we headed out for an *apéro*. We stopped in a random store to buy a white button-down shirt for Wildman, which he put on right away; he also wore the olive-green blazer we'd purchased together in Scotland. Holding hands and chatting, we walked toward La Cave à Michel, the tiny, standing-room-only bar where Wildman and I had quenched our thirst together on his first visit to see me in Paris, over a year earlier. Rounding the corner, we noted the Vespa outside the bar, signaling that it was open.

The owners, Romain and Ioulia, perked up visibly as we entered. Before long, we were deep into a bottle of something light red, aromatic, and zippy. Romain quietly put out a few hard-boiled eggs and a small dish of mayonnaise. Wildman ordered a second bottle, and an hour later, we were rushing off to the Bastille, late for the opening of *La traviata*. We had to watch the first act from the foyer, on a television. Finally, we were ushered to our seats. Wildman explained bits of the plot to me in whispers, though I was happy to sit back and absorb the climbing and sinking pitch of the voices.

And I was content to feel Wildman's warmth in the seat beside me through the performance, and later as we ate mediocre *steak frites* at a random brasserie, and then at night, as we lay in bed together for the first time in months.

The next evening, I had booked us a table for dinner at Jones, a small bistro on a quiet street not far from Gaba's apartment. We had eaten at Jones before, and on a separate occasion Wildman held a tasting for the Parisian wine community there. It was the kind of place we loved because the cooking was original, sourced from the freshest and most seasonal ingredients; yet the floor staff was unassuming—it was not a Michelin star–aspirational environment. It boasted an excellent selection of natural wine. We often discovered something we'd never tried before.

Before dinner, we met Gaba and Dan for a glass of orange wine at La Cave Paul Bert. Rory napped peacefully in his stroller in the corner while we chatted about the baby's sleeping schedule. They headed home, and Wildman and I made our way to Jones on foot, holding hands in the mystical, energizing Paris evening chill.

About halfway through our bottle of Romuald Valot's supple Beaujolais, the server delivered a plate of seared quail to the small table. Wildman began telling me about some news that had come to him via Giorgio de Maria, the importer in Sydney.

Like many importers, Giorgio was very close with the producers whose wines he imported to Australia, and he was very concerned about what had recently happened to the Sardinian winegrower Gianfranco Manca, who had a natural wine brand called Panevino. In Sardinia, Wildman told me, a devastatingly rainy season had destroyed most of the island's grape harvest. Panevino made zero wine that year, and Gianfranco was distraught.

"This guy's entire harvest was completely ruined," said Wildman. "Forty-five days of rain, in the springtime, so the vines couldn't even produce flowers. Hard to imagine, eh?"

Knowing that natural winemakers live vintage to vintage, with no safety net, not making any wine would be financially devastating for Gianfranco. It was also likely that he would suffer emotionally, not having the chance to reap the year's harvest after putting so much energy into his farming.

"Giorgio invited us to Sardinia," continued Wildman, slicing into the browned bird's breast and forking at the celeriac remoulade alongside it, "to see if there's something we can do to help. I was thinking we could ask Gianfranco to make wine with us in Australia. To recuperate some income."

I sipped the Beaujolais. "That's a great idea. The Panevino wines are incredible. We used to sell them at the shop I worked at in Brooklyn." Wildman's eyes were alight with excitement. "And I have always wanted to go to Sardinia."

He pressed me. "So that's a yes?"

But then I reconsidered. Leaving Paris to visit Sardinia was ideal for Wildman, who felt restless after four days in a city. In the fourteen months I'd been getting to know him, I observed that he lived to help others—he'd found many ways to assist upstart winemakers back in Australia. He believed that natural wine existed on this foundation of mutual support. It was a kind of global community, keeping each other afloat in the name of better farming, better wine, an ethical way of living and consuming.

I admired this impulse to help. But traipsing off to Sardinia at a moment's notice equated to more stressful air travel, less time in Paris, and missing out on my dance classes. I had just finalized plans for a launch party for *Pipette* at a Belleville bistro and was eager to make sure it happened without a hitch.

"So, just like that, we're off to Sardinia? For what, three days? Sounds kind of exhausting, to be honest."

Wildman blinked at my disgruntled tone. "We don't have to go. Or you don't have to join me if you don't feel like it." Gloom descended over our table.

I sighed. "It's not that I don't want to, but you know, I've been living here. I have plans. You didn't even think to ask me ahead of time."

"Well, I'm asking you now." Wildman hadn't taken a bite of the quail but now he threw back his wine and refilled it, not bothering with my glass. Obviously, he'd expected me to be thrilled about Sardinia. "It won't be all business," he continued, determined. "We'll have time on our own there."

Just like the first instance when Wildman had come to see me in Paris, I had to make a choice. It was between my persistent desire to be independent and alone, and my adoration of this city, on one hand, and on the other, the feeling of companionship I felt with this man, the knowledge that he loved me, that we could build something together, if I could let go of my obsession with Paris.

I loved the city's diversity, its thriving natural wine scene, the literary history and architecture. I had become close to a few people there, and of course, there was Gaba, who was like a sister to me. But it did not seem likely that Gaba and I would be opening a wine bar anytime soon since both of our lives had become unexpectedly complicated.

It did, however, seem very likely that Wildman had a question he had posed to me once before, which he wanted to ask me anew—and I suspected he would be doing it in Sardinia.

At the Cagliari airport, the espresso cost one euro and was entirely perfect—just the right balance of bitterness and sweet notes, with a creamy meniscus. We drank it quickly, eager to get into our rental car and make the one-hour drive to Nurri. The hilly landscape of short, green shrubs and sandy, compacted soils flew by outside the car window. I observed it with tired eyes from the early-morning trek to a remote airport outside Paris to catch our budget airline flight, while alert with the same appreciation I always felt for seeing a new place.

All I knew about Panevino was three things: one, that Gianfranco made an incredibly enticing skin-contact white wine called "Alvas," which consisted of about eleven different grape varieties, most of them unique to Sardinia; two, that the winery name related to the family's history of being the village bread maker and wine supplier for generations; and three, that Gianfranco rarely traveled, as he was not much of a "scenester" and cared not at all for marketing.

Wildman was driving with an expression of great contentment. Was it because he was used to living at the end of the world that he was thrilled to be on this remote island?

"I guess my wines are safely on the way to the US?" I asked Wildman. We had prepared a pallet of Persephone wines for shipment just before I'd left for Slovenia.

He nodded. "They should be arriving at the port in about six weeks."

Thinking of friends in New York, Portland, and elsewhere around the country being able to drink wines that I had harvested, fermented, pressed, bottled, and labeled myself in Australia gave me a shiver of delight. We'd also shipped wines around Australia to various shops and restaurants in Sydney, Melbourne, and Hobart, as well as to the few natural wine spots in Adelaide.

Wildman had helped make all that happen.

"Thank you," I said. "So much."

"You did it, you made the wines," he replied. "And they're really great. You have excellent winemaker instincts."

"Hardly," I replied. But I appreciated the compliment.

In a short time, the map on Wildman's phone told us we had arrived in the village of Nurri. We pulled into a driveway, where a tall, brawny man with dark brown eyes, thick black hair, and gray beard stubble emerged from a one-level stone house. Gianfranco blinked at us in the cloudy afternoon light until the car stopped, then opened my door and grinned. As I stood, he kissed me warmly on both cheeks, then embraced Wildman the same way. We inhaled the sweet, warm air.

"Let me take your bag," said Gianfranco in English, smiling.

He waved us toward the back entrance to the house just as an elderly woman with a slightly hunched back and gray hair pulled into a bun emerged from the front door. She, presumably Gianfranco's mother, was carrying a small, dead piglet over her shoulder. Her slate-colored eyes caught mine but she didn't stop, and we paused before rounding the corner to watch as she brought the pig over to a stone structure surrounded by stacks of wood and, holding it carefully with both arms outstretched, placed it inside.

Wildman and I exchanged a look—this was going to be a very special visit.

– Fifteen –

Appeasing the Rain Gods

"Ciao, ciao ciao ciaooooooo!" A chorus of greetings rang out as Wildman and Giorgio embraced, and then I took my turn kissing both cheeks with our red-headed importer friend. Giorgio had been in Italy for weeks already, visiting producers (and his mother, in Piedmont), and he radiated happiness from being in his home culture. A few other people were on the patio in the shade outside Gianfranco's house: a guy with salt-and-pepper hair and stylish, round eyeglasses sitting in front of a MacBook at a long wooden table, and a school-age boy who was playing with homemade clay-and-stick figurines on the floor.

"You made it!" Giorgio went on to explain that he considered Panevino sort of his second home, he came here so often. I knew he was close with all his producers, but this relationship seemed even more special.

Giorgio gestured to the guy with the laptop. "This is Gianluca," he said, and I connected the dots—Gianluca was none other than the natural wine pop artist, known simply as "My Poster Sucks." His brush-stroke pen artwork comprised the logos for many natural wine restaurants and events, including Giorgio's own Rootstock. We shook hands with Gianluca, who said he lived in Turin, a small city in Italy's north, but loved visiting producers with Giorgio.

Wildman and I sat down at the long table, and Gianfranco, who had disappeared into the house, presented us with a plate of hard sheep's cheese and salty prosciutto strips alongside slices of fresh bread.

I silently repeated their names in my head: Giorgio, Gianluca, Gianfranco. All we needed was a Gianmarco for things to get really confusing. I figured I'd probably make a mistake at some point and stopped worrying about it, diving into the bread and cheese.

"That bread is made from my family's four-hundred-year-old starter," Gianfranco told me in his clearest Italian—since I spoke Spanish from living in Spain and South America in my early twenties, and now French, I could understand Italian fairly well.

The bread was soft and chewy, with just enough bite to the crust. "It's so delicious," I said.

The wrinkles around Gianfranco's eyes bunched up when he smiled with pride. This was a producer whose wines were known worldwide, yet he rarely left Sardinia and had probably never traveled outside Europe. He was obviously someone who valued his home, family, and local culture.

It's tempting to forget how young Italy is, because the cultures that constitute it as a modern nation are so incredibly old. But up until as recently as 1861, when the grand Risorgimento campaign unified Italia, Sardinia was part of the kingdom of Piedmont. Prior to that, it was under the house of Savoy, during which time Sardinians defeated an attempted French invasion and revolted against the feudalist system. Their fierce spirit was eventually repressed by legislation that officially privatized land ownership, which had previously been collective in the tradition of the local Nuragic culture. This distinct linguistic group is documented as far back as 1800 BCE. Today, Sardinians—Gianfranco and his family included—speak primarily Sardo, a distinct language with Catalan and Northern Italian influences and Latin at its core.

Gianfranco's older son Isacco, a twentysomething with a passion for fine art painting, appeared wearing round eyeglasses and baggy sweatpants. He looked like someone more likely to be found in a Brooklyn hip-hop nightclub than a remote island covered in vineyards and sheep-grazing meadows. Giorgio explained that Isacco painted the Panevino wine labels, which changed year to year. We also met Helena, Gianfranco's wife, a woman with short gray hair who exuded warmth and calmness.

Needing to use the bathroom, I stepped inside the house, finding myself in the kitchen. And there was Gianfranco's mother, the *nonna*, diligently rolling out dough on a counter. Her hands moved so swiftly, I felt she could probably do this with her eyes closed.

When I returned outside, I asked Giorgio what the nonna was making.

"That's a local specialty called *culurzones*," Giorgio explained—they were to be stuffed with potato, pecorino cheese, and herbs.

Then two more rental cars arrived—one bearing a Japanese wine professional who hoped to procure some Panevino for his soon-to-open Tokyo shop, and another with a trio of French mother and daughters, including a sommelier in the London Michelin-starred restaurant world. Neither Wildman nor I was surprised to be joined by more wine professionals. In the natural wine community, visiting producers via appointment is vital in developing relationships and ensuring transparency of farming and production. Wine buyers and importers, after visiting producers, return home to share photos and stories with their teams, so that they could better understand (and sell) the wines.

Before the afternoon light could fade, Gianfranco ushered everyone back into their cars and led the way out toward the hills we could see in the distance. After being briefly detained while a troupe of wooly sheep strolled calmly across the road, we drove on toward the land where Gianfranco cultivated wine grapes.

"Have you ever seen vineyards like these?" Wildman murmured as we hiked up a slope, the entire international lot of us.

I shook my head. These high-elevation, rocky vineyards planted with hearty bush vines, meaning they were not trellised with wires, were very unusual. The closest thing I could think of was in the Roussillon, in Southern France, where bush vines grow near the sea amongst large slabs of gray schist. Wildman had worked at wineries in Oregon, New Zealand, South Australia, and Germany, but he was very impressed by the Sardinian landscape.

Each of the Panevino vineyards was unique. A few had almost no visible topsoil. Vines stood proudly above brown schist chunks that crunched under my boots as we walked. In some places, olive trees were scattered among the grapevines.

"Some of these vineyards are as much as one hundred years old," Gianfranco explained as we crowded around him.

These small plots had belonged in his family for a few generations, although they suffered years of neglect at one point. In the 1980s, Gianfranco began taking over the family vineyards and slowly rehabilitated them through careful tending by hand (the slopes were too hilly for any tractor). The native Sardinian grape varieties planted had names I'd never heard of: in one site, Gianfranco mentioned Monica, a red grape. In another he pointed to the white Italian variety Moscato, which I was familiar with. As well, there was Cannonau—the Sardinian version of the red grape Grenache, the most cultivated variety on the island. One vineyard, planted with Monica and Cannonau, was called Pikadé, and it was usually vinified and bottled as its own wine. However, Gianfranco never made the same wines from one vintage to the next, except for that skin-contact white I loved, "Alvas." He waited to see what the vineyards professed to him at harvest time and determined the year's blends based on quality and taste. It was more or less how Wildman made his wines, I reflected.

Other than occasional doses of copper and sulfur, considered organic compounds, not a single chemical had been used in recent years in the vineyards—no weed killer, no pesticide, no nonorganic fertilizer. In some years, Gianfranco didn't treat the vines at all, a rare thing for even a natural winemaker. No sulfites were added to any of the Panevino wines.

Gianfranco said he was too sensitive to harsh chemicals and he was also morally against the notion of treating his plants and wines with adulterants, Giorgio translated for us. It was clear that many, many hours of handwork had gone into keeping these vineyards pruned and tidy—and now, all for nothing. To have forty-five days of rainfall in May and June, just when young buds are shooting, preparing to develop berries. A year's harvest, totally unrealized. No treatment of any kind could reverse the damage done by that much rain.

Back at the house, the long table on the patio was set for fifteen, with a wine glass at every seat. The sun was lowering, but my light denim jacket was enough to keep me warm. I sat beside Wildman, across from Giorgio and Gianluca. Just before we dug into the potato ravioli, someone called out for a toast.

Across the table I saw Gianfranco, who must have been exhausted and hungry from showing us around, cringe slightly at this call. We were now obliged to spend two minutes reaching our glasses around the table and enforcing the perennial "rule of eye contact" as we clinked glasses before, finally, sipping the bright orange "Alvas." The Sardinian sun had provided ample alcohol in the wine. It tasted of ripe tangerines and spicy lemongrass, and when I swished it with a bite of the food, the pecorino in the ravioli sharpened immensely on my tongue. I used a piece of bread to mop up the oil on my plate.

Gianfranco's mother did not join us for dinner. I asked about her.

"She prefers to dine upstairs," Giorgio told me. I imagined the nonna enjoying a moment of peace and quiet on a well-loved couch by an open window, hopefully with a nice pour of wine, after her hard work.

Following several hours of drinking, chatting, and enjoying the roast suckling pig we'd seen earlier, Wildman and I said goodbye to the Japanese visitor and the group from London, who departed for nearby hotels. Giorgio and Gianluca went inside the house to some guest bedrooms, and Wildman and I were shown to a lovely, wooden guest house a stone's throw from the patio, where we slept deeply.

We awoke late the next morning, noting in the daylight that our little shack was set amongst an olive grove and a large, sprawling oak tree, and wandered toward the patio outside the house.

Gianfranco greeted us with a tray of espresso and toast, and told us that Giorgio had raced off early to catch a flight, taking Gianluca with him. But, he said, we could stay as long as we liked. His English wasn't bad at all, and he also spoke excellent French, or sometimes if he spoke slowly in Italian I could understand.

"If you don't have plans today," he said, "I propose you join Helena and me for a trip to visit an archaeological site that is very important in Sardinian history." We were game for this excursion. Gianfranco went inside to get ready.

"I'm so glad," said Wildman as we drank the last of our espressos, "that we will get to see something other than vineyards." I completely agreed—sometimes, when visiting a wine region, it felt myopic, to be incessantly discussing the soil type or land ownership history, without learning as well about cultural history, architecture, or politics.

Then Wildman said, "Maybe later we can take a drive, see the ocean." And I remembered—we weren't here for business only.

The four of us piled into Gianfranco's sedan and drove for about an hour through softly sloped, moss-green hills—many planted with vineyards. As he drove, Gianfranco explained that we'd be visiting a water temple from antiquity, where the Nuragic people of Sardinia had practiced rituals to bring much-needed rains.

We walked around the archaeological site. Gianfranco pointed out how the structures were used: there was a well for drawing water, a few towers, a temple. The irony wasn't lost on me, that in a year when nonstop rain had devastated his harvests, Gianfranco was showing us a temple to the rain gods. Perhaps he was hoping that his ancestors could do something to appease those gods, for better harvests to come.

Helena was quiet, listening. She told Wildman and me via Gianfranco's translation that as she was from Campania, she was always learning about Sardinian culture, even after decades of living there. The four of us finally tired of walking and peering at the stone structures. We sat on a bench and propped up my camera on a rock to take a lopsided photo of us all.

I wondered if Gianfranco would really venture all the way to Australia to make wine at Wildman's farm. More importantly, I wondered if I would be returning to Australia. Squeezing Wildman's hand in the back of the car as we drove back, I couldn't really see myself moving on from this relationship. With his poetry-infused language of living, his dedication to being an ethical person and setting an example through winemaking and farming, and the strong embrace he held me in through the night, I had to admit: this was what I'd been looking for. I did love Paris, of course—Paris was easy to love. But a city couldn't love me back. It couldn't make a family and home with me.

Back at the house, we had a late lunch of snails cooked in a tomato sauce and some freshly grilled local fish. Wildman asked where the nearest beach was, and Gianfranco gave us directions. Then his entire family went to have a nap, and although I would have liked one, too, Wildman and I grabbed jackets, jumped in our rental car, and headed east, toward the coast.

Back in South Australia, on days when the sun forced its way through the winter clouds, Wildman and I occasionally walked along the beach outside Adelaide, often with a bottle of wine in tow. It was picturesque: a long stretch of fine sand, gentle blue waves, and a rugged coastline that revealed Australia's ancient soils. Once or twice back in the summer, we snorkeled in a cove there. Flipping our feet under the water, we held hands and watched squid changing colors and an octopus hiding amid the rocks.

It was on that South Australian beach in late July, as we walked Alfie and Lulu in the shallows of the tide, where I had ended our quasi-engagement, which Wildman had surprised me with just before hitting "play" on *Barbarella*. The isolation and frigidity of the Australian winter had taken an unexpected toll on me mentally. I felt confused about our relationship, including the rapid pace at which it had evolved over the last year, I explained. I needed to go away to Europe, and I wasn't sure for how long.

"So, do you not love me then?" Wildman had asked me in a horribly pained tone. I struggled to come up with a sincere response. I did love him—didn't I? But I wasn't sure what I wanted. I felt like I was being a shitty person, but I could not deny my restlessness and uncertainty.

Now in Sardinia, we again beheld sandy coastline, stretching for miles, and a few locals walking their dogs. As we walked across the damp sand in silence, I thought back on that day when I broke up with Wildman. Was it just the gloom of rainy, cold winter, coupled with my lack of close friends and the difficulty of driving in the Hills that had caused my hesitancy? If I went back to Australia, I wondered if I could push past these obstacles. I'd made it in New York for nearly eight years. Had managed to get a French visa and built something of a life for myself in Paris. I could do it again in Australia if I wanted to.

It would be worth it, I felt, given how special this relationship was. It's not every day you meet a South African–Australian winemaker in Georgia when you're planning to move to France, and he woos you in

Edinburgh and Slovenia and Spain, and you make wine and honey together and raise puppies together and start building a home together on a stunning farm where tropical birds and sheep coexist.

It wasn't every day that someone loved me as much as Wildman did.

Wildman was squeezing my hand. Each step felt like I was lifting my foot from quicksand. All my senses were heightened.

"Are you cold?"

I nodded sharply, and Wildman took off his thick beige jacket whose sleeves were covered in oil smears from tractor work, and put it over me. We continued along in silence for a few minutes, and then he cleared his throat. I stopped and turned, and saw utter vulnerability in his eyes.

Wildman lowered down on one knee. He reached into his pocket, and there it was, in a small box: the rose gold band adorned with a blue sapphire. It was the exact ring we'd commissioned months earlier, after the Tasmania engagement.

"Will you marry me?" Wildman was asking. I looked down at his soft sea-colored eyes, his sturdy shoulders, the shiny ring in its box.

I reached out and took it, and put it on my finger.

In the word "yes" is contained multitudes—it's an agreement, a joining of oneself to an idea, a future state of things. "I will" means "there is a future." And it's us. It's your visions, your life's work, merging with mine; it's a dialectic narrowing its focus, becoming a synced conversation instead of two individuals facing apart from each other, yelling out into the universe.

It is also, by nature of exclusion, a "no." Yes means no, I will not be a perennial wildflower, growing wherever I please. No, I won't make decisions in isolation. No, I cannot put myself and only myself first anymore. No, I won't be living in a studio apartment in Belleville, writing in silence, eating noodles in a Vietnamese café at midday for the rest of my life.

Yes, I will risk giving up my hard-won independence to be with you. Or rather, I will find a way to incorporate that independence into a partnership. I say yes in order to bring our two worlds together, in hopes of creating something stronger, more tangible, more everlasting, than whatever we could do alone.

With Wildman driving, we headed back inland and slowed the car in a town that appeared to have one lingerie store, a post office beside a bank, a fuel station—and surely somewhere, an old Catholic church. It also, fortunately for us, had a bar. We went inside and ordered negronis, which came with a dish of peanuts, then carried it all to a table on the deck overlooking the quiet street. It began to rain as we shoveled handfuls of peanuts into our mouths and drank our bitter negronis quickly.

"Don't you think we should have the wedding sooner, rather than later?" I asked, looking down at my ringed finger. I was suddenly aware that we'd opened a new act in our relationship, in our lives.

"Yeah," Wildman replied. "Keep it simple. Cozy."

"So, maybe just after vintage." I reached for the pouch of tobacco we were sharing and began rolling a cigarette.

"The weather is good that time of year." Wildman lit the cigarette for me.

The rain slowed to a drizzle, and I inhaled its scent as it hit the sidewalk, mingled with the tobacco smoke rising above our table.

"Another round of negronis?"

"That sounds good." He got up and went to the bar.

I watched him disappear inside, then stared at the empty street and the downpour while he retrieved the drinks. For once, the rain didn't spoil my mood.

Wildman and I arrived back at Gianfranco's family house at dusk, just as Helena and Isacco were setting the table for dinner.

"We have some news!" Wildman grabbed my hand and gleefully held it up for our Sardinian hosts to view. I flushed with surprise and swatted at him, embarrassed by being paraded around.

The family congratulated us, and I relaxed about it as we sat down to eat. Gianfranco poured us one of his red wines, and the rich, full-bodied Cannonau warmed my blood.

Wildman began convincing Gianfranco that coming to Australia to make wine at the Lucy Margaux Farm was a realistic plan. Apparently, Giorgio had already discussed it extensively with Gianfranco. Wildman was making it clear now that it would be possible to find extra organic grapes, and that Gianfranco would be not just accommodated but welcomed, treated like a VIP.

Eating and drinking slowly, Gianfranco listened attentively. He seemed to be deeply considering the idea. I was impressed by Wildman's dedication to the concept of a Panevino guest residency in Basket Range. He didn't just talk big about mutual support in the natural wine world—he actually made it happen.

After three nights in Sardinia, we kissed our host family goodbye and said we looked forward to having them in Australia for the coming vintage, then headed back to Paris. We were just in time for the launch of *Pipette*'s first issue at Le Cadoret, a natural wine–focused bistro in Belleville.

It was a warm autumn evening, and I wore a lightweight black dress that hung loosely over my body like a big T-shirt and boots that put a spring in my step. Wildman had brought over a few bottles of Persephone in his suitcase. As the event started, I stood by a table spread with pâté and fresh bread and crudités, beside a stack of magazines. As the guests arrived—writers and photographers who had contributed to the magazine, as well as readers—I poured small tastes of my Gamay and Sangiovese. I hadn't tried them since bottling day, months earlier.

A Swiss friend who had a popular Instagram account full of eclectic natural wines paused after a drink of the Sangiovese.

"Wow," he said. "That's . . . really good, Rach."

Though I didn't say it aloud, I agreed—both of the wines were truly beautiful. Maybe just because I had made them? It was hard to be objective, of course, but they both had a silky, pure quality, with acidity perfectly matched to the fruitiness of the Gamay, and with dusty tannins atop a bouquet of fresh violets characterizing the Sangiovese. I flashed back to the little green shed on the hill overlooking the veggie patch and muscling that lever on the basket press to squeeze the juice out of the grapes.

We opened a few more bottles of wine as the crowd grew. But where was Gaba? I'd been expecting her to walk in with Rory in a stroller early on. Just as we were finishing up, the text came from her: she'd somehow lost her wallet, or it had been stolen. She sounded frantic and upset. I shrugged it off but also wished she'd been there.

"It was the one time she could have tasted my wines," I said to Wildman after we thanked the bistro owners and headed out to join people at a nearby wine bar.

"I'm sorry," he said, kissing me on the cheek. "But all those other people got to try your wines. And how many magazines did you sell?"

I shook off the disappointment. But Gaba's absence seemed ominous. I somehow felt that she would never taste Persephone. That our friendship, spawned in New York but transplanted overseas, was destined to only exist within the confines of Paris. With a ring on my finger, I was finally headed in the direction of building a family. But I had to admit that this path would be divergent from, not coinciding with, my closest friend in the world.

That night, we returned to the apartment around 1 a.m. in a taxi from our last stop on the natural wine trail and went directly to sleep. It had been an exhausting week, going back and forth to Sardinia and hosting the event for *Pipette*.

Mid-REM, I was rustled awake by the glow of Wildman's telephone. His face, illuminated by the screen, revealed distress.

"Sweetie, what's wrong?"

He blinked a few times. "It's the shipment to Europe." This referred to a few pallets of Lucy Margaux wine from the recent vintage, which were currently stuck at the port of arrival, Le Havre, coincidentally not too far from Paris. Sighing deeply, Wildman explained that there was a holdup because of some certificate he'd sent in too late, which apparently was extremely vital. Now, he had to send emails and probably express-post the certificate, or else the wines would be moved to a storage facility, and he'd be charged something like €500 per day.

I rubbed his chest. "Well, you can send it off in the morning. There's no point in obsessing about it now." He didn't seem to hear me. I got up and went to the kitchen, where I had a small bottle of whiskey that had comforted me during many insomniac nights over the past few months.

Back in bed, I passed Wildman a tumbler and sipped my own. He continued to swipe and press on his screen for a moment, and then he dropped the phone in exasperation.

"I'm also feeling pressure because it turns out the divorce can't be finalized until the land sale goes through."

I let that sink in. Wildman had long been legally separated—"I'm single, baby," he liked to assure me happily whenever I'd inquired about the situation. But he wasn't divorced, officially. There was a property realignment under way, which needed to be completed first.

We chatted about this for a moment, and Wildman's worried tone softened as he got it off his chest. As we emptied our tumblers, he seemed ready to fall back asleep and laid an arm over my body, stroking my hip softly.

There was something I'd wanted to ask for a long time.

I whispered, "Did you still love her? When she left you?"

Why did I need to know such a thing? I was trying to gauge the emotional trauma that Wildman would have experienced, having a fifteen-year marriage come to an abrupt end. It happened mere months before he and I met. I wanted to make sure I understood what I was getting into. What were the depths of the original pain? Had he had a chance to heal?

Ideally, we would have discussed all of this months earlier. But we'd been neglecting these emotional realities, instead busying ourselves with travel and winemaking. We had filled our days with work and romance to avoid confronting some of the challenges we might face.

After a beat, Wildman replied, looking up at the ceiling. "Yes, I did. I did love her. It came apart because I was working too much and I didn't put time into my home life. And she never really said anything, just made plans to leave." He spoke drowsily, as if just uttering those words were a release that might allow him to return to sleep.

It was the first time I'd heard him speak in such an unguarded way about the separation. A mix of emotions came over me—first, the realization that Wildman had his heart truly broken and then fear that I might always live in the shadow of their relationship. But at the same time, I felt much better hearing him confess that he'd loved and been hurt.

It didn't quite put us in the same situation. Our relationship histories could not have been more different. But I was now assured that we were both recovering from dysfunctional romances, of very different sorts. Mine was a masochist obsession with New York City. I leaned in and kissed Wildman, and he stirred out of his near-sleep, embracing me and pulling my body toward him.

In mid-November, I said *à bientôt* to Paris—I refused, of course, to say the more permanent *au revoir* to the city, or to Gaba, or to anyone who lived there. I insisted I would be back, as soon as possible. As if to prove this, I tucked my tall leather boots, a few dresses, and my collection of Mavis Gallant stories into a box and shoved it under Gaba's bed, promising I'd reclaim them "next time." It was symbolic but it made both of us feel better. With a kiss to her, Rory, and Dan, I then lugged my bursting suitcase to Charles de Gaulle Airport. I knew the way so well, I didn't even need to check the route.

After sleeping a few hours on the overnight flight, I awoke as the plane was landing outside Washington, DC. It had been nearly one year since I was in the US. I didn't miss certain things—the news cycle, for instance—but now I felt warmth inside at the prospect of reconnecting with my family. If there was one thing the Signer clan did well, it was Thanksgiving.

Three days later, my mother and I drove back to the airport to pick up Wildman. I was, of course, nervous. What would my mother, a woman whose career in reproductive rights advocacy had involved working in an office not far from the White House, think of my farmer fiancé and his wardrobe of old T-shirts? Would anyone in my family raise an eyebrow at the age difference?

Once he'd retrieved his bags and kissed me warmly, Wildman shook hands with my mom. "I'm Marj," she said kindly. "How was your flight?" To my relief, they chatted briskly all the way back to the house.

And then the man I was going to marry was sitting on the couch in my teenage bedroom. My mother was getting ready to move out of the house where she and I had lived since I'd been in high school, so, thankfully, I had spent the past few days scrubbing away all the hideous fan posters I had plastered all over the walls at the age of seventeen—numerous Grateful Dead signages, a Dave Matthews Band album cover, and a concert poster for Outkast.

That night, we planned to have a quiet dinner and let Wildman catch up on sleep. But my two older sisters and older brother, upon hearing that he was there, arrived with takeout from the Caribbean Grill, a fast-food place that had enlivened many a weeknight during my high school years. They wasted no time in grilling him.

"So, you have a farm? As in, with animals?" Asked my sister Becky, a lawyer.

"You're from South Africa? Wow, what was that like? I'd love to bring my sons to Africa one day," said my brother, Mike, also a lawyer.

"Did you really make this wine? It's amazing," exclaimed my sister Mira, who worked in the mental health field, staring into the glass of Pinot she'd just been drinking.

"I sure did," laughed Wildman. "But wait until you taste hers, it's really good," he said, nudging me. I said we'd have it at Thanksgiving dinner.

Wildman was clearly exhausted, having come all the way from Australia, but he jumped up to help with the dishes.

"That accent is adorable," Becky murmured with an approving smile after he'd whisked away our plates. They all seemed to love him. The next morning, Wildman and my mom delved into a heated discussion about American politics (her favorite topic) while I stood nearby, drinking coffee and mentally saying goodbye to the house.

My mother and Wildman carried on and on. She and I always wound up snapping at each other if we spent more than half an hour together. But they were enjoying each other's company. Finally, I broke them up—we had to get ourselves to the festivities.

Thanksgiving for years had been held at Becky's house on a tree-lined suburban street. We crowded into the kitchen and I started making a round of Campari spritzes, while Becky opened the oven to show us the turkey she had been roasting since that morning.

"Becky, wow," I said. Perhaps I had been away from the US for a while and forgot how dramatic the hormone-pumped animals could be.

But my mother agreed—it was the largest turkey any of us had ever seen.

"Are you sure that's not a deer?" Wildman joked as I handed him a drink.

Hours later, it was nearly time to eat, and Becky was frazzled, unsure about whether the turkey was ready. Someone called out, "Look up the cooking time!"

"I would do that," she murmured to me. But she couldn't remember how much the bird-monster weighed.

I shrugged. "Ask Anton," I said. "He studied cooking."

Becky gasped. "Please help us!"

My fiancé was now in charge of this high-stakes oven situation. He gave me a look but opened the oven and poked around the bird's rear end to see if any blood came out.

"This is done," he pronounced. There was applause.

I went to the fridge and brought out a bottle of my Persephone Wines Gamay. The label, made of watercolor paper, was just starting to peel around the edges, but otherwise it looked great.

"I made this," I said, pouring a small glass for each of my family members. I watched my mother, who notably was a fan of Beaujolais, take a sip. Her eyes widened and she said, "Wowwwww."

Everyone praised my Gamay. Nevertheless, I had the feeling that my family didn't *really* believe I'd made it. I almost couldn't believe it myself—there in my sister's dining room, with the people who'd known me all my life to be a bookish girl who maybe liked to party a bit too much, I was having a hard time imagining myself in that green shed overlooking the veggie patch, jumping on grapes. But that was what I had done—it was who I was becoming. One day, they would all see. For the time being, the Australia that I was returning to in a week was like a secret that Wildman and I shared.

After a few days in Baltimore and New York, hosting events like we loved to do, we took a train to Brooklyn for a wine-soaked dinner at the institution Roberta's Pizza with Chris Terrell, the importer who had brought us together in Georgia and who now brought our wines into the US. Then, Wildman and I boarded the long-haul plane back to Adelaide via a layover in Melbourne.

We were headed back to the dogs—would they remember me? Back to the springtime, veering on summer (I'd cheated the seasons again!). Back to the veggie patch, to long meals with local bottles at the Aristologist. Back to that mess of a house on that mess of a farm, where there was always so much to tidy and improve and plant and organize (next, we would deal with the kitchen, which was desperately in need of renovation). Back to sunset drinks at Manon Farm, spooning Monique's homemade ricotta onto her freshly baked sourdough bread, savoring their spectacularly tart Savagnin amongst the vines they tended so carefully.

Back to the view out of the tall French windows in the bedroom, where I woke a few days later to fog rising over the eucalyptus-covered hills above the valleys. A pair of broad and sturdy wedge-tailed eagles circled in the sky, and I watched them, mesmerized by the hourglass figures they seemed to be drawing in the glow of the rising sun.

Wildman came in, Alfie and Lulu at his side. The dogs jumped into the bed, and he handed me a milky coffee, just the way I liked it. I pointed out the birds.

"Beautiful, aren't they," he said, adding, "They mate for life."

And then he headed out the back door, down to the winery, leaving me in bed with Alfie and Lulu, who were snuggling gratefully at my side, as if I'd never left at all.

Epilogue

For the 2019 vintage, our Sardinian friends Gianfranco and his son Isacco came to Australia to make wine. It was astonishing, seeing Gianfranco tasting grapes in the vineyard, making split-second wine-making decisions in a country he'd never researched or visited. Giorgio, our importer friend, came over and made giant vats of *paccheri* with lamb ragu, using our homemade passata. Giovanni, the owner of 10 William Street, came to help with picking and processing grapes. Aaron, from the Aristologist, joined the Lucy Margaux team officially that year.

Surrounded by Italian men, many of whose names started with *G*, there I was, making my second vintage of Persephone. Friends from around the Hills came to help me with the basket press, and I made five red wines, one Chardonnay, and one pét-nat that year. The Chardonnay did go a bit volatile, and it was a lesson in humility—sometimes, a naturally made wine isn't perfect.

Shortly after that vintage, a month before our wedding, Anton (a.k.a. Wildman) and I learned I was pregnant. It was our little secret as we said our vows in front of a small crowd on the beach near Port Willunga, our snorkeling and walking spot.

In December of that year, Simone was born. When she was three months old, I made my third vintage of Persephone Wines with her alongside me in the shed. While I foot-stomped grapes, operated the basket press, and checked on fermenting barrels, I rocked her in the pram or even held her in my arms. The rosé I made this year, a coferment of Merlot and Sémillon, I named Cuvée Simone for the little girl I can barely keep up with.

Winemaking continues be both awe inspiring and at moments—especially in the colder months—a difficult life. We never know what the vineyards will yield and whether it will be sufficient. Disgorging sparkling wine in the wintertime will never be my idea of fun. And Australian life has its challenges: there were raging bushfires last summer, which wiped out many vineyards nearby.

Anton has recently planted several vineyards on the hills around the Lucy Margaux Farm with the cuttings we took over my first winter in Australia, as well as more taken the next year. After driving the tractor up and down those hills to plow the earth, each vine was planted by hand. As I worked on this manuscript, Anton and Aaron were using a hand hoe to pull out the gnarly weeds that grew over the winter of 2020. We just began thinning the shoots on the new vines.

Now it's spring. Anton and I take walks around the property with Simone on our backs in a baby backpack. She is enthralled by the birdsong and the hanging eucalyptus canopies. We take her into the veggie patch and let her chew on calendula petals or crawl around looking for ripe strawberries while we harvest our dinner.

Over in Paris, Gaba and her friend Lorna finally opened Tuesday Addams. The place launched as a pop-up, serving Gaba's signature cocktails and plates of tandoori octopus, with glasses of natural wine. Unfortunately, the coronavirus hastened a swift end to Tuesday Addams. But I have no doubt that by the time we finally make it back to Paris, the bar will be up and running again. By then, Rory will be four years old. He and Simone, who will be nearly two, can chase each other around the playground while Gaba and I sit on a bench, having croissants, letting the flakes cover our jeans. Once we've tired the kids out, we'll bring them over to Septime La Cave to meet Dan and Anton, who despite being so different, always manage to make conversation. The kids will sleep on their dads' laps while we decide which bottle to order.

"Hm, what about an Italian wine, something from the South?" Gaba will say.

"A skin-contact white?" I'll propose.

"Here's this pét-nat by La Sorga, I haven't had one of their wines in forever."

"Is that Danish natural wine? I've heard good things."

"I think I see La Garagista on the shelf, over there."

The possibilities are endless. If it doesn't exist, I guess we can just make it ourselves.

Photo by Sally Wilson

Acknowledgments

THANK YOU TO MY AGENT, LAURA NOLAN, WHO IMMEDIATELY SAW the potential in this book and stuck with it through various phases. Our relationship was what I always sought in an agent—she was caring and devoted with rigorous intellectual discussion and the critical eye my proposal required. The fact that she loves natural wine was a wonderful bonus.

It was our luck to have Lauren Marino at Hachette Books come on board as editor; she edited with a swift, confident hand, exactly what I needed. Thanks to others at Hachette who diligently copyedited and thoughtfully designed this book.

I am grateful to my husband, Anton, for always cheering me on, through the publication of each issue of *Pipette* and the writing of this book, and for always being a willing debate partner or fact-checker when it comes to the subject of ethics in natural wine or extremely detailed aspects of viticulture. Thank you for encouraging and supporting me to make wine, something I never anticipated I would do.

To every individual who has read a copy of *Pipette* or drunk a bottle of Persephone Wines, and to the retailers, thank you for keeping these projects going. Each time one of you sent messages, saying that you loved the magazine or the wine, they touched me deeply. Thank you to everyone who has worked on *Pipette* over the years.

In New York, I found an indispensable community of writers and mentors in the CUNY Writers' Institute Program. Thanks especially to the writers who gathered regularly at Karen de Luca Stephens's house to scrutinize each other's work—and especially to Karen for hosting. I also had a wonderful community of wine lovers in my biweekly tasting group, and I thank all of them as well as Clara Dalzell, who hosted many late, raucous nights.

Thanks to all the individuals who lived the scenes described in this book, including those who taught me difficult lessons, but most profoundly to the ones who lent me a couch or have extended their friendship beyond borders and over the years. And of course, to you, Gaba, for being you: fearless and incredibly human.

To my parents, Marjorie and Robert, for raising me in a house full of great books and insisting on editing my English essays in high school before I handed them in, and again to them as well as my siblings, Mira, Becky, and Mike, for emotional support throughout the years as I found my way toward a writing career, thank you all.